Inward ⊙ Way

by
AJ Karlovich

"Life is an expression of our inward characters - a spiritual blend of art, poetry, and motion."

Introduction

To begin any ponderance or contemplation of our cosmology, and how one fits into that picture, it's pertinent to examine our dreams and wishes. The very idea that our minds create our reality, is quite remarkable when we allow ourselves to delve into it. Therefore, by abandoning social norms and cutting ourselves some slack, we're bound to experience some pretty remarkable things.

For this work, we'll begin with a question, and further, the thoughts and feelings derived from this question, and see where they lead us. Aristotle brilliantly stated, "philosophy begins in wonder." Well let's see if those wise words from the wisest of material minded thinkers is accurate. So without further delay, our challenge is presented...

What kind of world do you wish to live in?

How does this questions make you feel? Does it fill you with wonder or do you draw a blank? You see, in truth, the wondrous feelings that emanate from this kind of question, will set the tone for not just the rest of this book, but further onward - your life.

As a result, it's our wonder which perplexes us, and makes us want to unite with forces beyond ourselves, this planet, the solar system, and this galaxy - all the way to the Universe as a whole... What does It feel? Does It, or, has It felt the same as us? Are we sharing the same emotional dialogue? It's no surprise that a human being would want to associate their thoughts and feelings with a universal measure. After all, haven't we all at some time or another come to the conclusion that we're in a relationship with the greater cosmos? It's an eccentric fantasy that pulls us in many abstract and intangible directions - for the purpose of seeking union with a much larger cosmic family. Furthermore, because we live with this kind of ponderance nearly all of our lives, we would have to question whether the Universe is undergoing the same process that we are?

This Universe, which is one of reflection, is the ultimate contemplation, where even the smallest most minute things are seeking to "know thyself." Whether they be microorganisms or planetary bodies, all forms of creation appear to be, for one - either trying to find meaning, or two - trying to find their place. And somehow - someway, we live with the funny feeling that the Universe is discovering itself through each of these things, including us!

So in a sense, the faint echoes of universal wonder have trickled into our imaginations, sending our thoughts far and wide. In a way, the cosmology that we create from this meditation, is what formulates not only our place in our galaxy and solar system, but also how we view our relationship with this planet. That's why our original question is the most important a human being can ask itself right now, at this particular moment in time. In fact, when has it not been an important question to ask? When all's said and done, doesn't our view of the world keep the dreamer dreaming, and the fool - wishing?

Now by fool I don't mean the modern version who's looked down upon, but the fool of old who thought beyond their reason. This fool who innocently pondered more than he or she could conceptualize, often found themselves swimming in dilemmas. Well that should spark our curiosity... Shouldn't we all, in some way, be considered this kind of fool? Aren't we all wishing and dreaming for a better world - a world in which we can freely be the highest actualization of ourselves? In turn, aren't we all swimming in dilemmas? From this inquiry we find a rather peculiar thing... the fool who continues onward in question, eventually graduates their ignorance and becomes the philosopher.

By one measure they've become an anomaly, and by another - a pioneer. They're no longer indebted to societies reason, as most are. In point of fact, the philosopher is one who froths at the idea of pushing the envelope and dreaming bigger. In light of this, we should all be fools in our own right!

Thereupon, when we let go, and allow the dreamer to dream - a philosopher will emerge. Simultaneously it seems, our observations lead us toward something that's even more apparent, but no longer explored to it's fullest extent... Life is quite strange - it's a paradox of endless expression!

As our inner philosopher continues to dive deeper into our dreams and wishes, eventually wisdom endows us with the ability to transcend our boundaries of thought and feeling. In time, our preconceived realities lose their appeal - forcing us to reconsider our beliefs. Therefore it's our wishing that starts us out on the path of discovery - one that keeps us asking for more and more.

So when we ask the familiar question "who am I?" or even "why am I?" - in some way, shape, or form - we're expressing our divine wishes. This gives an interesting twist to our dilemma, and makes it more of an idea we constantly ponder. It can perhaps be considered a background noise to our conscious attention - ever-present in some way. Ironically, we find this query or idea intertwining with not only the concept of our world, but who we personally wish to be or become, and furthermore - how that person fits into this world? Respectively, our hopes and wishes because of this conceptual entanglement, directly relates to how we see ourselves.

Who are we, really?

To answer such a magnanimous question, we must dissolve the paradigmal cloud we've built around ourselves, and allow the unknown to ripple. When we allow the intangible and unthinkable to enter into our thought-waves, more and more possibilities begin to emerge. What does that tell us about the current state of affairs in the world today? Well, for one, it says that there's an unknown amount of solutions that might exist for our outward dilemmas. And two, they exist beyond our inner barriers, which we've been living to protect.

Until a person breaks down these walls, and frees up space to allow the originality of the greater cosmos to flow through them, they're simply living a preconceived reality, or, as a thought in form.

We have to recognize that all forms of thought that exist in this paradigm today, are no different than the forms or archetypes that existed yesterday. Which means there's little originality or uniqueness in our current concepts - or at least how we perceive them. As a result, the answer to our original question is greatly limited. And so it becomes prevalent that we must further examine this unique experience in order to truly understand what life is trying to show us.

To begin, where does one look to find the answer to this question?

Well, primarily, we find it within and without ourselves, which is a clever expression or way of saying, "with imagination." You see, we're all gifted with a mind that can search the infinite possibilities, and in turn extrapolate logical opinions from an endless cache' of wisdom. However, we must understand that each opinion is only a solution to one level of being. Which means, there's not just one, but many levels to our true inner characters!

This leads us to the second step, which is equally as important as the formulation of the opinion - When we do find an answer, we have to let it go and explore new possibilities, with an even larger or deeper understanding of the original question! For example, if it's left to you to formulate an opinion of the world, you must be able to have an answer to the question, "why me?" and then, "who am I?"

Our questions therefore continue to ripple further into the realm of rising knowledge, which means the variables that are explored, are done so in a thorough manner - at as many levels as the individual can imagine. This may sound like the path to theoretical madness, or even an inherently flawed program based on infinite contradictions. Yet that's the key to the nature of this game we call "humanity." The possibilities are infinite on the one hand, and "in finite" on the other. This means, our true nature, as well as inherent purpose exists within us, or, within our "finite details." Which also tells us that our organisms hold not only the secrets to this world, but all of the cosmos.

And so this contradictory game that began with a question, continues onward with questions - creating the kind of environment where we aren't stuck in any particular mindset, or thought form. Hence, both our pursuit and understanding keep evolving.

By allowing our imaginations to continue creating in their own unique way, we as individuals become the outlet for Creation to continue creating within and without itself. You - the one who asks the question, are the only one who can decide the answer or riddle to your own life. Your will is free(free-will), and therefore you can become anything that you wish to become. From this realization, you must know that your life is conceivably limitless in respects to the intangible realm of thought and feeling. So let these forces cascade!

By the nature of our source, which is none other than divine creativity, we must continue on working with the same form of thought that makes creativity so special - It never decides that any one thing is final! It always continues onward and inward in a state of perpetual change, which is the nature of the cosmos. Creativity by this point of view, is a state of being that's constantly changing - allowing our thoughts to spiral into new and uncharted territories. It matters not your individual classification, or whether you choose to call yourself a person of creativity or of change. The bottomline is - if you can grasp the idea that collective identification is limiting, you're choosing the path of uniqueness... This entryway is where every one thing is recognized as the Universe focused into one point - existing in a state of discovering and rediscovering Itself. It's a contradictory nature that denies the finality of the thought-form, and allows our minds to evolve with the eternal expression of the "Now principle." Meaning, each moment in our lives is an expression of One Moment's continuance. Moreover, we must understand this One Moment as an experience that's not static or stationary - It's something that remains in motion and continues to change, because it's the manifestation of endless creativity! You must be free to evolve your thinking in a manner that coincides with this expression. By doing so, you'll eventually realize the creative mind's ability to continue evolving it's greater view and understanding of the world. Which brings us to the idea that our role here is a responsibility, but even more, an incredible gift! We're called "the measure of the entire Universe," because of a divine curiosity - almost as if it's our mission to continue measuring and evolving the Universe's measure of itself. We do this by creating the greatest possible reality for our individual lives, here, on this remarkable planet we call "Earth," which is an anagram for "heart." Thus the Earth is the heart of this wonderful experiment we call, "Life!"

Chapter 1

The Nature of You

In an experience of many different frequencies, resonating to an infinite level of vibration, there remains a singularity we're all seeking to merge with. Call it God, Life, the Universe, Creation, etc. Yet there's something remarkably odd about this search that must be addressed... If we're living within the experience we call the Cosmos or the realm of Creation, why is it we feel as though we're a separate organism - excluded from it? It's as if all the beauty we witness through the windows we call eyes, are taunts or reminders of what we feel, not. But, you see, that's the illusion, and it must be in order for us to find meaning in Life! All those wonderful things we witness outside ourselves - all the beauty and creativity, even the drama and chaos, is all but a reflection to who we are on the inside. Our inward character is the outward beauty and chaos of the creative world, and vice versa. Now, perhaps you're feeling a bit overwhelmed by this, and still experiencing distance from uniqueness. So why don't we dive into the "why" question that eats at us incessantly.

A human being - heralded as the becoming "measure of all things," should be no stranger to this Universe of frequency. So why are we?

Well, we start by looking at how the majority of us live this experience. And when we take a good look, we find humanity living in the midst of a philosophical struggle - one that's been around since our earliest developed memories. It's that age old paradoxical question that envelops us all at some point -"Is life a game of materiality, or, is it a game of spirituality?" We've also referenced this as Science vs. Religion, or Body vs. Spirit. In fact, this struggle in one way or another, is merely the representation of the two polar opposites of being - positive and negative, strength and weakness, higher and lower. The same debate collapses on the wave-particle duality, or whether an electron is a wave or a particle. Is life about energy or is it about matter? Time or Space? Relativity would tell us that life is a combination of both - an equation if you will. However when researched closely, we find by the simple observation of the atomic world, life is 99+ percent energy and roughly 0.00001 percent matter. This goes for all existence in the Universe. Even the emptiness we see in space, which we once thought a vacuum, is now said to be teeming with charged particles. This fact, though foreign to the material minded person, is a truth that many choose to live unaware of. And one has to wonder why that is? Especially when it appears to be so beautifully obvious and extraordinarily creative!

To further complicate the issue, being that much of what we don't know exists in a field of energy that we cannot see, we choose to rely mainly on what our 5 senses(sight, sound, smell, taste, touch) can pick up. This is the mindset of the faithless, or, those who are not open to trust the Universe. Now before we move forward, it is prevalent to separate what one might call faith on the one hand, and belief on the other, because they're often confused as being the same.

The word belief comes from two root words, which are "be" and "lief." "Be," as we can all guess comes from "being," which is "a state of existence or living expression." "Lief" comes from the indo-european word "leubh", which means - "Love." So the literal translation of the word belief is "to be in Love with" or "what you Love," and is generally used as a matter of deep opinion. For instance, belief denotes a person's idea that the Universe will turn out in such and such a way - "I believe the Universe is endless," or by a worldly example - "I believe the Earth is the center of the universe," and so on. Faith on the other hand has many ideas and translations, which in my opinion have perverted it's true intention. Many use this word as a matter of strong belief or trust - typically in the religious sense. For example, someone may say, "my faith is in Buddhism" or "my faith is in Christianity." Others use the term in more of a direct way, as in putting their trust into someone or something. For example, "I put my faith in God or Allah" or "I have faith in Christ." Yet the words truest translation relates to more of a state of being, where it simply implies "openness," or "the experience of being open." This openness is more of an envelopment of and for, either letting go completely, or taking in the Universe entirely. It can also relate to a state of open heart and mind, where we put our full trust in life and the cosmos.

So in order for a person to be faithless, they would have to be closed off to any particular part of the Universe, usually because of narrow-mindedness. Unfortunately, this is where many of us are today - in need of change...

At some point in our history, we as a species decided to put our full trust in our beliefs, rather than live for the experience of openness as we once did. Because of this, we live only for what can be classified as real, or what can be physically recognized by our 5 senses. However, what the majority of us don't realize, is the tremendous amount our 5 senses close us off to. The Universe, as stated, is only 0.00001 percent matter or physical form. For this reason, because we're fixated on the material world, we're almost completely narrow-minded in our view of the cosmos.

When we live only by the physical nature, we're closing ourselves off to nearly everything in the Universe - living only as a conscious product within the mili-fraction of the existence we call "Life." Life, as we can decipher, is an anagram for "file," which expresses Creation's inherent desire to accumulate some kind of information - perhaps simply for the purpose of discovery. Therefore, stagnation or living for what has already been discovered - keeps us from adding to the file. Furthermore, because of this closure, we interpret our higher authorities under the same principles. After all, God or Allah is a physical being, right? Christ, Krishna, Dionysus and Osiris will all come back some day, won't they? Yet what we continue to miss - in all of our worshipped characters, is the Nature of what it means to be "the measure of the entire Universe," or... You! Through this character, which has been labeled under the extreme polarity of prodigy and imbecile, is where we truly discover information that can be added to the file(Life).

Unfortunately, since we live in a world that's almost entirely belief driven, our ideas of ourselves are limiting - greatly reducing the possibility of us measuring up to anything more than a flawed creature who suffers and sins away their potential. Be that as it may, when we drop belief, or loving an idea more than we do the freedom of expression, we can begin to explore and or discover ourselves completely. Through this freedom of expression, we can ask ourselves questions about our own nature and not fear reprisal from beliefs or ideas. After all, sooner or later we all ponder the question, "Who am I, really?" or "What is the nature of this being I take ownership of?" And from these questions we find an understanding that humanity is seeking to truly know their origin and purpose. How else have belief-based religions prospered?

When we open up and embrace the inward path, the questions begin to pile up, but our wisdom continues to rise. Wisdom, nonetheless, is the state of living with philosophy, which derives from the latin "philosophia" and the greek "philosophos." "Philo" meaning "to Love" or "to befriend," and "Sophia" is in reference to the Gnostic, Telestai, or Pagan Magna Mater(mother goddess) of this planet, who's name translates to "wisdom." So philosophy literally means - "A Love or befriending of Sophia(wisdom)." And as we can all agree - this planet is wisdom!

Through this living wisdom, we're able to find a more universal approach to our cosmology. But, more importantly, we're able to discover the truth of our anthropos(human species), which is what Creation seems to be requesting us to do. By recognizing this request, we're ultimately led to the challenge of all challenges, presented by the phrase of all phrases - "Know thyself."

To know ourselves, first it's wise for us to go to our most obvious source, which was called Sophia in the mysteries, or, as it is translated - wisdom; the very nature or spirit of this planet.

Well, at some point we've all been captivated by the miracle of nature, and wondered endlessly how it came to be? The mountains, fields, deserts, rivers, lakes and oceans - continue to capture our hearts, setting fire to the fuse we call passion. Eventually, snowballing into an eruption of imaginative brilliance.

To know the mysteries of this remarkable planet, we have to ask ourselves, "what is Nature?" What is this spectacular enigma we continue to shrug off time and time again? Is it just a dazzling phenomena of randomness brought on by the Universe's superior creativity? Or is there actually something tangible we can learn from it?

This universal flow that we typically perceive to be a physical paradox of not just this world, but all the cosmos, most often in our simplest description - "just is." And then the majority of us move on - not daring to dive deeper into the puzzle.

You see, this Nature; the "Tao" or "way" as it has been referred, is more than just the physical cause and effect of motion in life, but is also the very framework for the energy that makes all motion in the Universe possible. It's the formula for random beauty which interweaves all waves and frequencies that create our reality. Taoist refer to it in chinese as, "ziran," which refers to "spontaneity," or "something occurring naturally." So to say, "one must exist within their own nature," is also expressing one's desire to be in complete harmony with more than just their individual being, but also the natural functioning of the cosmos. The same goes for the outside environment, which continues to change moment to moment as if time were a mechanism, or perhaps a pulse to a revolving clockwork illusion. The tick tocks mirror a small image of the larger experience - it's interrelated.

This clock circles 360 degrees, letting us know, for one, that all things revolve in some form of pattern, and two, the revolutions are endless. Time will keep ticking as the Nature of the One Wave we call the Universe continues to ripple further onward and inward. Where we are going, what we are doing, is the result of this larger flow - this force of Life. So our participation in this Great Wave, often referred to as "Nature," is vital for us to know ourselves.

If we don't seek to further understand "Nature," we will remain trapped in the current fear paradigm that challenges our organism - which seems to always be operating at the mercy of its beliefs in the physical environment.

If this is the case, it means we must all individually explore nature in order to find meaning and understanding. Furthermore, it would appear that we also need something else to correlate truth. This correlation is found through a somewhat obvious question... "What is the Nature of a human being?"

We all know human beings have a physical nature, we're certainly aware of the mental, we claim to be in touch with our emotional, and we all wish to be in the grace of the spiritual. Wow, wait! Hold up! A human being has 4 different natures? And they're also called "the measure of the entire Universe?" Then that would mean the Universe is an accumulation of these same 4 nature's? The answer to these questions are all "yes," but there's more to it than that.

You see, considering we hold the beliefs of our existence in 4 separate parts (think of them as 4 waves of energy existing within us), the goal is to align all 4 parts to form one whole - one wave. What would we call that wave? We would call it - "Universal!" Therefore, a being that exists in harmony with the complete nature of this one wave, would be called a "Universal Being." And what would be the complete Nature of the Universe? The answer to that is... Creation.

Beyond any and all thought of what Life might be about, or what this Universe is seeking to achieve, Creation continues to be the driving force, as if everything in itself was a smaller version of the larger Creative Energy. Whoa! Stop right there! Everything is a smaller version of the larger Creative Energy? Well, of course! "Made in the Image of God" ring a bell? Every single one of us is a product of the One Image seeking to know itself in a particular form. It's form, for us, is the being we see in the mirror every day. But remember, everything created by this measure is unique! That's why no two people are exactly the same. There's not one eye in the world which is exactly the same as another eye. No stone is the same as any other stone. There's not one cloud which is or was exactly the same as another, and so on. What we eventually find is endless creativity! Furthermore, in order for the One Self or Image to know itself, IT must continue onward creating more and more versions of itself. Life's an expression in this way. It is an expression because of the act of creating, which is fueled by the emotions received when one discovers something they didn't know before - thus perpetuating the need to keep creating! A human being expresses itself in the same way. We are all - each of us, seeking to express ourselves fully in our own individual nature - with the hope of discovering uniqueness in the Universe! That's our purpose in life - To discover and then live as we are, without fear of reprisal for being anything but unique thought and form. However, knowing the nature of oneself is no small task, which is what makes Life such a wonderful experience.

So, where do we begin this journey?

If Life is a journey, we have to ask ourselves, "what does going on a journey require?" To know the answer to this question, we first need to become aware of Life's changing frequencies. A journey after all is a product of frequencies - it's a set of outward motions or wave intervals guiding us towards a destination. While riding this wave, we experience countless events or moments that bring us wisdom - furthering our knowledge of truth. And when that waterfall we call "death" finally arrives, our common hope is the achievement of inner knowingness, which also coincides with passing the test, or solving the riddle of our own lives(in order to get into heaven). It's the pot of gold at the end of the rainbow scenario, where mankind is expected to strike it rich - so they can move on into more luxurious accommodations. Perhaps a mansion with an indoor swimming pool on Angel Street, where a 1,000 virgins are waiting to fulfill our every desire. Or maybe it's a place with majestic gardens that smell as sweet as lilacs, and cloudless skies forever - except that which is below our feet. After all, isn't this how our material minds work?

Of course our beliefs in heaven are countered with the opposite, where if we don't pass the test - we're sent to a place that's an eternal dance of fear, chaos, and destruction, or, perhaps a hell where we suffer the same death over and over. This idea rivals the material mind's ability to believe that stagnation or purgatory is a literal place - one experienced as a result of a person's life.

Yet, all of our dreams and wishes eventually come up short when we begin to take a closer look at ourselves. After all, what good is wishing for a pot of gold if we're not moving onward in search of it(adding to the file)? Likewise, isn't motionlessness, or living within one standard revolving lifestyle, the very purgatory we're wishing to avoid?

It's these kinds of questions that border heresy in our religious model, which compels us to keep our heads down. They're also the questions that prod our curiosity - propelling us to take risks that are against the common grain. However, these questions, which are inherently perceived as rebellious to society, make us look twice at our current life path. They're the questions that make us wonder beyond normal reasoning with a fool's heart; a heart of hope.

So, we must understand that it's our wonder which makes us journey - whether it be for the inward or outward experiences. And just like the echoes from those philosophical giants, Socrates who sired Plato, who sired Aristotle - "philosophy begins in wonder." Therefore, so does every journey!

When the courage is found to explore the unknown or uncertain, the journeyman or women tends to venture outward in life. That pot of gold at the end of the rainbow becomes the final puzzle piece. But, it comes with another expectation... The same as answering the riddle of the Sphinx, one is expected to answer to God. However, your answer is not only expected to be in the same variety as a student giving a verbal dissertation, but also as a composer presenting their magnum opus, or, divine symphony. Many of us believe that God requires us to play this symphony of redemption during our judgment at the pearly gates in front of St. Peter. This desire to become perfect in one's expression or ownership of being, pushes the hopeful mind in a direction where all experiences become missing pieces to a grander puzzle.

For example, by a simple minded interpretation, if we don't have the job, we can't own the car. And if we don't have the car, we can't have the job. On the same hand, if we don't have the job, nor the car, we can't own the house in the suburbs with the two car garage. If we don't have either of these three, we won't have enough stability for companionship. Without companionship, we can't create the missing link that furthers the expansion of our species. And without children who will pass on the family name and legacy, our lives are without meaning...

It's these kinds of material minded pressures that keep our pursuits within the confines of a norm - they're a part of the designed program set forth by society. So our expectations in life are always a product of an outward game of divide and conquer, or separate and achieve - we continue to create separate arenas for our goals (ie. job/car, house/spouse, child/legacy, etc.).

Furthermore, it makes us believe that these material expectations are staples to what bring us Love. For example, "The spouse and children will bring me Love and a knowingness that I provided for the world. The house gives my Love shelter and security. And the job and car bring a means of comfort for the growing desires in the empire I've built on Happy Street."

Fortunately, at some point, we all begin to reject this model, which brings many of us to the dreaded midlife crisis(which is now happening at younger and younger ages). This is where the typical material minded slave upgrades all the above. Perhaps a new car to prop up their ego. Maybe a bigger house with a workshop or office to find more privacy from their dependents. Even more scandalous, maybe they'll pretend they're in Love with someone else to feel alive again. If we can make it through this era of temptation unscathed, perhaps we can actually take time for ourselves. This brings us to the only point in our adult lives where we're allowed to truly explore ourselves - retirement, which is synonymous with "old age." This of course is when we've built up enough credit and confidence to exit the game, or perhaps when the world has decided we're no longer of use.

Thus a person's individual exploration seems to be something that's only left for the child and the elderly - at the beginning and the end. Why else do so many of us live with regrets that stem from our youth? Isn't it because we've never figured ourselves out to begin with? Perhaps, just maybe, we blame others for our own misfortune of not knowing ourselves?

Today's youth recognizes the struggle of their predecessors, and like all new generations, have been rebelling. Yet rebelliousness today is also stuck in the material mindset.

The world today is based upon a rapidly growing technological society, which further develops, cradles, and micromanages life. Most of our youth are lost in this virtual world of instant gratification and the false idols of non-reality television, which are nothing more than the fabrications of yesterday's material concepts.

So does life ever really change for a society? Or does it maintain the same game we discussed before - material versus spiritual? Or outward versus inward? Our lives seemed to always be left up in the air, and given to us as a basic contemplation - "where and how will our individual natures develop in societies game?"

What it seems like we're continuing to miss, again and again, is a truth the majority of us eventually discover, but rarely embrace... "One's own nature never begins or ends outside of oneself." In fact, it is merely you, right now, right where you are - inhaling and exhaling. You are your own nature, which is the uniqueness! Which also means... that pot of gold we're all searching for, actually is the person that we've always been...

You see, we may be able to find some clues as to who we are by observing our environment, but we'll never truly know ourselves until we observe our being in motion, right now! That is why, in my opinion, humanities true philosophy should neither be a material philosophy nor a spiritual philosophy, but more of an Inward Philosophy - an Inward Way.

When we go inward with Life, we're better able to understand, reflect upon, and express ourselves with the outward. This Inward Way is an approach, as well as a path, where we examine the Nature of all things, both inside and out, and craft them into one process, within us. This form of comparison is an expression of being that's not what many may perceive - collapsing into or isolating oneself - not at all! It's about allowing the 4 energies that you are, to begin to be as they are - an expression of one complete wave, which is You!

The Inward Way is a path to Universal thinking, feeling, and being. In this state, your mind, body, and spirit, are fully expressed through the higher vibrating emotional language that exists within, without, between, and of all things. This is where humanity is progressing towards, even in the midst of societal dismay, which appears to be a natural expression in the cycle of life - "In order for a human being to know the best of itself, it must experience the worst of itself."

So today the majority of us live in a material world which is completely compartmentalized and separated by beliefs that vary from culture to culture. And of course, these compartmentalized cultures - drowning in separatism, are bound to clash. This clash of beliefs, spawns from a collective mindset that refuses to look inward. Thus we've created a world with many layers and puzzle pieces that don't quite line up or fit. As we've continued to force this puzzle together, our basic structure has grown to a point of extreme instability - a society ready to fall apart completely.

But, what we must understand, is that humanity - like the individual, is experiencing a wave which must complete its full intervals. Civilizations rise and fall, only to rebuild stronger in the rising Sun. The night produces the day, and the day produces the night. Is night greater than day, and day > night? Of course not! Polarities are designed for us to experience life's duality - it's changing intervals. It's beneficial to know both sides to an equation in order to know the purpose of the question. Our answers in life come from the wisdom gained in observing these polarities. So we must evolve ourselves to become just like that - a wave that continues onward completing its intervals; changing with the tides of life; the revolving night and day.

In order to know oneself or "the One Self," you must become the inward and outward night and day, the rising and falling waves of the ocean, and even more, the symbol of Nature's endless flow - A Creator in the flesh... It's quite the challenge we've been given, isn't it?

Chapter 2

A School for Spiritual Progression

The Soul is a vast source of mental and emotional energy that accumulates our individual identities. It's not only our Spirit's personality, but also that of our physical body's. However, considering the soul is the compilation of many lifetimes of experience, knowledge and karma, we, in one life, do not have conscious access to all of it. Basically our current lives are only but a small representation of who we are spiritually, not only for the purpose of focused growth, but for our body's safety. If our body, in most cases, was created with full consciousness, it's circuitry would overload - most likely killing us. This is why a person's waking up period is gradual. It gives our body's time to adjust to what we learn. For instance, it may be possible, and there's much evidence to suggest, that before we come into a physical life, our path is designed so that we face and incrementally awaken to certain aspects of our soul energy. This is in reference to karmic lessons, which incidentally only refers to causality of action. So basically we learn gradually by cycles of "cause → how will I respond this time?"

There's a strengthening case which dates back to early civilization, proclaiming the Earth to be a school for spiritual development. In fact, when it's observed in it's entirety, it's a rather challenging school considering what constitutes being human. Since we consist of mental, emotional, physical, and spiritual natures existing together at once, it would be like seeking your doctorate in everything as an infant. However, the real question is - what if the Earth was meant only for the advancing soul? Or perhaps soul's approaching undergraduate or graduate level education - rather than elementary school? This is plausible considering that the Earth before the turn of the 20th century housed roughly 1 billion humans. However, with the introduction of oil as an energy resource and processing material, in 100 years the population has spiked to over 7 billion. That's 6 billion new enrollments into this school, many of which are very young souls that were probably not ready for the challenge. So the Earth in a spiritual sense has become rather chaotic and dominated by a lower vibrating and more physically minded population. This is not to say that a young soul cannot be intelligent, or with the right guidance or path advance quickly. It just means that the nature of the school has changed dramatically in a short period. It has now become an extremely difficult school to graduate! Imagine seeking your undergraduate, graduate, or doctorate - in everything, surrounded by over-zealous and rather young children? It would be hard to get your work done, but that doesn't mean there isn't plenty of fun to be had(who has more fun than children?).

Let's take into consideration, that because of this massive influx of young souls, the collective consciousness has become chaotically imbalanced. Well, just like what we find in nature - balance, or, harmony is something which is naturally sought. And so, in order to counter this influx, much older Soul's(maybe even soul's who've already graduated) have begun to or have been incarnating to help the planet along in it's process. After all, a growing university needs teachers doesn't it? On the other hand, from the technological perspective, once we pass a certain marker and become a galactic species, we cannot enter space or make relations with much older or even younger races without a much higher understanding of ourselves. Imagine telling an advanced alien race who've overcome all the challenges that we're facing, and then some, that they must accept Jesus Christ into their heart, and furthermore "get baptized?" Or more jovially, imagine telling them to take their shoes off before entering your home? These are some examples of societal norms and pressures that must be either addressed or eliminated in order to seek more harmony for our species. Pressures and norms are the direct result of festering inner-conflict. There's largely too much chaos surrounding our inner-characters, which reflects heavily upon society as a whole. So in order for humanity to graduate into a galactic species, we would have to overcome these dilemmas. In addition, one would have to venture a guess that young warring societies are incapable of surviving in the new frontier, which appears to be geared toward inner-balance. Therefore, if we brought war to the stars, as well as technological irresponsibility, we would for one not be respected as we could be, and two, be highly vulnerable to making the largest of mistakes - resulting in the failure of the anthropos.

In order to combat the carelessness that follows the majority, it's essential that we slow down our outward progress and begin to explore our inner-dramatis personae or characters. This is the requirement for spiritual advancement, and theoretically what this school's original purpose seems to have been. Therefore, by reverting back to the original model and purpose of cultivating the Inward Self, undoubtedly it will help us solve many of our outward dilemmas.

It's estimated that roughly 20-30 percent of the global population is of the advancing or higher advanced soul level. Some believe this is enough for us to make a change in direction. However, when studying other species in Nature, take a herd of elk for example - it takes a 51 percent majority for the herd to reach critical mass, which is required for a decision to be made. Case in point, if the herd needs water, the lead female(not the male) will look toward the direction of the watering hole. Once this happens, one by one, the entire herd begins to look in the direction of the watering hole. When it reaches a 51 percent majority or critical mass, the entire herd moves to get water. This is natural democracy in action, but rather at the level of the "collective consciousness." So in human society, if 80 percent of the population believes that we're on the road to WW3, well, then technically we are! This is how every group mind operates - through a collective consciousness. If 80 percent of the world believes armageddon is approaching, then it will most likely happen. Now i'm not attributing this so much to the law of attraction(like attracts like), which so many of us carelessly use today, because it's purely an electromagnetic principle. When we collectively pull something to us, it's happening magnetically. All it takes to turn a magnetic force into a reality is electricity, which quantum physics proves is nothing short of awareness. So once we become aware of something, we begin initiating it into reality. Take an electron for example. When scientists seek to observe an electron - it's a particle. When they're not observing it - it's a wave. This means that consciousness directly affects the outcome of the physical environment at the most basic level. In this way, when it comes to any particular idea or event, as the awareness of that idea or event rises in the collective consciousness, the greater the possibility becomes for that idea or event's manifestation into reality. That's why Einstein said, "I don't believe in quantum physics, because I believe the moon is there even when I'm not looking at it." This of course was his way of jovially stating, "I can't believe awareness or applied consciousness manifests reality!"

So if the majority of our world leaders are of a younger soul frequency, which makes them more physically minded and emotionally undisciplined - because we're following their lead, the world will experience more physical and emotional conflict. This is what gives us a Warring Society. War however, seems to be fundamental for a human being's development, because it's a growing principle(by way of physical conflict). Yet at some point, because all waves rise and fall to become their opposite, the warlike soul changes its direction from an outward physical warrior, into an inward spiritual warrior. The difference between the two is great, because the Inward Warrior battles their own dilemmas and replaces inner conflict with wisdom. They prove themselves with discipline and using physical force as a last resort, until eventually they graduate their outward physical desires. This, believe it or not, is the transition that society is going through and must go through. We need to understand, first and foremost - our physical lives are only but a character we choose to play in order to advance ourselves spiritually. As a result, life is a form of education, built upon Creation's marvelous ability to dream many dreams, and make each dream seem more real than the dreamer. Physical life by this understanding is the illusion, and the spiritual life, the reality.

Let us look closer at this concept of Spiritual warfare. For one, we have to understand that the nature of war is in a polarity with peace. This is why many people of influence use the contradictory phrase, "War is Peace." To a young soul, peace becomes the result of war by means of physical conquest. Thus the United Nations goes to war in the name of peace. This is a rather ignorant view of the concept, perhaps even a bit diabolical, which many will tend to agree. To the spiritual warrior, war brings peace because the soul in question recognizes that their energies, or 4 natures, are at war with each other - basically they're out of balance. So in order for them to realign, they must go to war with themselves. However, in spite of this declaration of war, they're not attacking themselves with negative intent, but more of a desire to understand the cause of their imbalance. Thereupon the spiritual warrior becomes a student of life, and seeks to gain inner balance through knowledge, truth, and discipline - all for the purpose of achieving not just fluidity, but a direction of flow. This means, the desire to know one's self is for the purpose of finding one's own source of inner spirit, and learning how to use it wisely. Now when I say "use" I don't mean "control." By "use," I mean "co-opting" or harmonizing with one's own Nature - so that their actions become unconsciously in rhythm with Life.

The Spiritual Warrior has an inward philosophy, or, as I prefer to call it, an "Inward Way." Their life is not for the purpose of material gamesmanship, even if the material world appears to have it's benefits. The reason why the spiritual warrior distances his or her self from the material, is to learn how to discern and respect their body's needs and wants. They recognize they are not their body, because the body is the vessel or temple that houses them through their period of education. Therefore, they seek to know and Love their bodies unconditionally, because they're responsible for it. This means, and as the old saying goes, "what you put into your body - you get back." The body to the spiritual warrior is the outward most expression of their inner-self. It's the beginning trials of self-discipline. So the more we learn to understand and respect our bodies, the more we're also respecting ourselves spiritually, because both body and spirit are in polarity. That's why some say, "The body is the spirit, and the spirit is the body." This of course is not to be taken literally, but more to reflect the work that's being done on the mental and emotional level. Which means - the work we put into ourselves directly affects the nature of the whole experience. So the body encompasses the entire experience, but we're still not the body.

The Spirit sparks the bodies creation, and as the body develops in the womb, the soul aspects meld with the heart, gut, and cranial brains. This creates the personality type as well as the challenges we must face both physically and emotionally. Yet, we're still not the body, because it's an illusion created by the spirit. We know this now, thanks to quantum physics, because energy does not become matter until awareness or consciousness is applied to it. As a deduction, we can perceive the spirit to be the electrical awareness that turns energy into matter. So from that spark inside the womb, which started as a magnetic attraction, the spirit brought life to the body by first bringing life to the cells. This means - the cells that create us are highly intelligent!

For example, one strand of a double helix of DNA is responsible for roughly 125,000 different processes! All of these intelligent cells, which are created not only from the 23 chromosomes each from mom and dad, but also from the electromagnetic attraction-awareness of the spiritual world, co-exist to form the micro-universe we call the "human being." Even more, this intelligence is self-sustaining, which means we don't have to be conscious of how our cells and molecules remain together while we're in motion. Case in point - we can walk down the street without us worrying about our bodies falling apart! The same goes for any other movement or process in the body - it's all relative.

The spiritual warrior acknowledges the wonder and complexity which makes them the most highly advanced organic structure in the Universe. Furthermore, they know that it's also their birthright to be "the measure of all things." After all, they're using a micro-universe to measure the macro-universe! So if they cannot respect their bodies, they cannot respect their truest self, which is genius in every sense of the word.

There are many schools of thought, both public and secret, that believe in everything I've just described. However, the conflicts that arise between each of these schools or systems of thought are purely products of Ego, or, "concepts of self." This is how and why a religion like Christianity can have 10,000 different denominations seeking the same truth, but at the same time each claiming they're the only one who knows the truth. This is also how wars continue to plague much of the world in the name of a loving patriarchal God, who's about as loving as the inquisition was to Europeans. Or, by a direct example, how Loving this God was to Abraham when he requested the sacrifice of his son Isaac as a testament to his loyalty. It's also how science and religion are always at odds with one another to prove who has authority and who doesn't. You see, if we're not intuned with an Inward Way, and approach this challenge of being human, which has been called "the measure of the entire Universe," from the perspective of a spiritual warrior - all ideas, philosophies, cultures and so forth, will continue to clash until we destroy the anthropos completely. This is why we must reign in our sense of outward progress and turn our focus Inward.

The inward approach allows the individual, as well as society, the opportunity to achieve a greater knowing of Self, as well as a much greater relationship with the Earth(translates back to Sophia or "wisdom," but also is an anagram for "heart"). Unfortunately, this would require a complete overhaul - a paradigm shift.

This change in direction, nevertheless, is not something that can not be achieved at the societal level alone, because all great changes begin with the individual. Ultimately, it's you and I who have the freedom to explore our own uniqueness by means of an Inward Way. And by this way, our living examples make us lanterns in the dark - where eventually... we all light up the world.

Thereupon in keeping with our origins, which developed from a complete state of openness or faith, the planet will become a school yet again. It will happen, because the truth found through our Individual natures has always, and will always prevail. The eternal path that exists within all things will be re-lit for us. When this day comes, the anthropos will never shine brighter.

Chapter 3

By Way of Initiation

Society today is run by many forms of ritualistic initiations. Whether it be men or women seeking to gain access to social structures like fraternities or sororities, or the countless other cultural groups and orders that exist today - initiation remains at the forefront of our society. I don't want to say that initiation plagues us, but alternately its purpose has been perverted. The simplistic view of initiation has become more of a means of acceptance, rather than what its initial intentions represented. When we look back in history at the concept of initiation, we see it used as a tool, to again, either gain access to social or cultural groups, but also to be inducted into an education that was geared toward a higher wisdom. One example of initiation that's commonly known is in religious institutions, where an initiates first order of duty is to devote worship to a deity or divine concept. An example of this is when a catholic monk enters a monastery, offers his complete devotion to the church, and through ritual seeks to attain the grace of God. Another example is found in buddhism, where a person enters a temple, gives up their earthly possessions, and begins the path of desirelessness for the purpose of reaching Nirvana. These initiatory ideas are designed for the objective fulfillment of one's own concept of divine enlightenment, or, an emergence with a higher form of being. They're very ritualistic, in the sense that, each initiate dedicates themselves to certain disciplines in order to build the subconscious reflex that revolves around "letting go." This idea of letting go or "surrendering," is when one puts aside their own ego, or concept of self, and seeks to draw in or attract God's guidance for their development as a spiritual being. Yet when we delve further into the concept of initiation, and trace back its roots to our ancient past, we rarely find desire or reason for associating or merging with a divine presence that exists outside of oneself. The letting go or surrendering aspect was appropriately geared toward the removal of past beliefs, in order to initiate oneself into their original purpose, or, the idea that they are their own individual seeking to "know thyself."

To reflect further on this idea of "self-initiatory" practice, we look to the mystery schools and Pagan religions of old, which promoted a heavy emphasis on going inward to discover one's own inner mysteries and relationship to the divine cosmology. By this idea of divine cosmology and one's own inner mysteries, an initiate devoted his or her life to the path of wisdom, which was developed by studying the fundamentals of nature. Through their research and application, they eventually altered their life approach by following natural principles, which brought unity to their inner-being. Further, this growing relationship between one's inward self and the mysteries of the natural world, would bring the initiate cosmological wisdom. And by cosmological wisdom, I am referring to knowledge of truth, which is an understanding of the Universe in one's own light; their own way. Each and every individual by this measure would learn to apply themselves in a way that would make them a benefit for the development of not just the anthropos(human species), but also the evolution of the planetary consciousness.

The planet itself, or Sophia as it has been referenced in the Gnostic sacred ecology, is a divine energy or torrent manifested, which originated from the pleroma or galactic center of the Milky Way. Within this center is where the human genome was developed as a divine experiment. And the question of course was and still remains, "How will the anthropos measure itself to the Universe?" As a result, our species has been referred to as "the measure of all things" or "the measure of the entire Universe." This idea or concept traces far back into our history, and although distant to our current understanding, still holds a dimlit candle for the original hope and wish that our species would fully come into itself. Most of all, that it wouldn't be fooled by the trap of good vs. evil. Correspondingly in Judeo-Christian theology, we find two trees in the Garden of Eden - "the tree of the knowledge of good and evil" and "the tree of life or immortality." What we can deduct from this mythology, which can also be found in many other philosophies and religions, is that humanity has chosen the path of polarity over the path of divinity, which is why we suffer error or sin as it's called. However, we must realize that chosen paths, one way or another, are a measure of how something is meant to come into itself. It's a hard path we've chosen, because of the constant struggle between the idea that two forces are seeking to recruit us. At the same time, each side requires our allegiance or worship. This unfortunately is a cosmology that many are oppressed by, which has kept us locked in a perpetual polarity of "which side will win our souls in the war to end all wars."

This notion of good vs. evil was not the original intention of the experiment. In fact, it was considered an anomalous after effect that the Gnostics and many of the Pagan religions claimed was unintended. The Gnostic sacred cosmology, which resembles modern day science fiction, teaches that Sophia in her fall from the galactic center accidentally created a predecessor to the human being, which was void of many qualities that make the anthropos so special. Unfortunately, this anomaly, or unintended creation, was formed with a unique ability to exist as something inorganic - more of a phantasm in the mental realm. It was also supremely rational minded(think of a being with a 300+ I.Q.), as well as inferior to divine human emotion. This species that they've referred to as "archons," might be considered alien to our current understanding - having the appearance of an underdeveloped fetus, but also as a tall lengthy creature, which is similar to the large and small alien gray in our science fiction. Yet to the Gnostics, they were part of the beginning of the solar system, created before humanity, and thus have accidentally become an integral part of this experiment. Their conflict is one of being given tremendous ability in their infancy, and furthermore, robbed of the process of evolution, which allows a primitive form of energy to grow into a being of true universal intelligence. This outside influence, which has existed since the beginning of our cosmology, is found in many cultures around the world. In ancient

Greece, these are the titans and olympians of old. In Judeo-Christian theology, these can be considered the watchers, nephilim or elohim by one interpretation, or the fallen angels by the other. This idea of otherworldly or godly influence, and perhaps interference, seems to be inherent in our mythologies around the world today. And unfortunately, we've not developed enough intellectual savvy and perhaps courage to challenge our political and religious authorities into agreeing that these entities are other worldly. On the other hand, it gives the intellectual mind ample information to develop their own reasoning as to what the intention of these outside forces actually represent.

In the Gnostic mythos, these archontic beings are harbored in the outer planets(Jupiter, Saturn, Uranus, and Neptune), and developed what may considered a holographic civilization built upon precise celestial mechanics. Now by celestial mechanics I'm referring to a system that can be explained mathematically. It may be complex mathematics to some degree, however it's still based upon formulas, which tends to be how the pragmatic mind operates. So when we look at the planet Saturn, which has been called the crown jewel of our solar system, we find a very interesting phenomenon that truly captures our imaginations. For example, Saturn's north pole forms into the shape of a six sided polygon, or as it is referred - "hexagon." "Saturn's hex" is a cloud formation that spins at nearly 60 degree angles, and seems quite mechanistic. In fact, it has been theorized that Saturn is multidimensional(several layers of geometry), because this hexagonal cloud formation spirals downward toward a central point existing inside the planet. What captivates people even more about the nature of this planet, is it's ring patterns. Encircling this gassy giant are several concentric circles of dust and rock revolving together. What makes it even more of a sight to see, is when the Sun's light illuminates these rings. During this time, the rings light up and produce what can be described as a halo around the planet, similar to the halos described around an angels head. Yet, for all it's amazement, Saturn in truth, is an inferior planet in comparison to the Earth. Saturn appears to be created by one small spec(largely used) of the mathematics found here on this planet. In fact, the greatest fascination with Saturn is found in the ratios for Pi and Phi, which are simply key staples in the math witnessed here in this solar system. Pi is used to express the ratio of the circumference to the diameter of a circle, where as Phi, or the "Golden Mean or Ratio," has been used to analyze the proportions of naturally occurring objects(ie. sunflower, hurricanes, tornados, the orbits of planets around the Sun, parts of the human body, DNA, etc.). Though Saturn expresses these ratios quite beautifully, there's no comparison between Saturn and the Earth, and the reason is quite simple... The Earth operates from a creative mathematical formula, which is evolving itself constantly, whereas Saturn does not. Saturn can be referred to as a basic scientific calculator, where as the Earth can be considered a quantum processor. They both are similar in regard to basic functioning and theory, however, the Earth is far superior in its complexity as well as expression of creativity. In fact, the Earth is the epitome of creativity. So we would have to disseminate, just by this very fact, that Saturn exists as merely a point of study, not of worship, which we'll get to later on this book.

Getting back to the Archons - they're led by a demiurge or grand leader that's considered tall and reptilian in appearance, with a serpent or horned dragons head and a lions body(resembles the sphinx?). His creation was part of the overlord class of this archontic species, and was a natural occurrence to the anomaly. However, this archontic overlord overstepped his bounds(a result of creation without evolution), and claimed to be more than he was. This fantastical interpretation by the Gnostics holds great resemblance to the horned deity found throughout our mythology, which can be associated with Cernunnos in the Celtic traditions, Cronus or Saturn in greco-roman mythology, Satan or the Devil in our western theologies, as well as many others found around the world. Yet, the Gnostics go further, and even boldly claim that this being, who is "god-like" in it's creation, has falsely claimed to be the divine creator of the Universe, and is actually a demented alien who has been referred to as Yaldabaoth, Yahweh or Jehovah - the male God worshiped by the Abrahamic religions. On the one hand, his psychological profile can be compared to Loki in Norse mythology, who operates as a grand deceiver, trickster or illusionist, and is more interested in playing human than actually being one. On the other hand, his demented approach towards Creation, which portrays supreme vengefulness towards his cousin counterpart who has been given the gift of the human genome, fits more appropriately with the Christian version of Satan - who pulls the wool over the eyes of mankind. Furthermore, this struggle between good and evil, is actually the fictitious and rather diabolical manipulation of this outside entity.

This archon, who realistically has no power over the anthropos besides psychic manipulation, is playing the leader on both sides of the coin. And further, he has manipulated the collective consciousness through a form of psychic terrorism, for the purpose of us creating the destructive path called armageddon or doomsday, where he plans to intervene as the savior...

This story may sound completely insane, but nonetheless it's what the Gnostics claimed. I guess now we can truly understand why the Gnostic teachings were subdued and scrubbed almost completely from our history books by the Catholic Church. It also gives us an even greater understanding as to why they've been the most persecuted group of people in our history - which is fact! Though at this point, I'm going to stop discussion on this story for two reasons. For one, it's not the purpose of this work. And two, there are far too many people who follow the belief that this patriarchal God is the master of their Universe. Yet, these people live in fantasia, because they're concept of him is based entirely on a belief system, which has been packaged nicely by religious institutions who've abused and manipulated humanity for thousands of years. These "believers" clearly have never met this self-proclaimed God, nor have they dared to explore the physical, spiritual, and psychological horrors he stands behind(the list is endless). But I will definitively say this... It is irrational to believe that our purpose in life is to bow down and surrender our free-will or life force to an outside entity in order to feel accepted as a human being. The notion that we're flawed creatures, who are born sinners, is a gross and rather disgusting idea that tarnishes our much older traditions, which envision the anthropos as something extraordinarily beautiful. Our own misdeeds in life, or true error, is continuing to believe that we're inferior to our own image - which is divine in every sense of the word. Because of the delusionary continuance of this inferiority complex, many of us remain blind to our true inner characters, which are screaming for us to wake up.

We all have a part to play in the anthropos, but even more a purpose in discovering our own individual nature. In the past, those seeking to "know thyself," initiated themselves into a higher wisdom in order to find unity in their way of being. Furthermore, this inner knowingness would also grant them the ability to be a vital instrument in their peers recovery of purpose. Hence the initiate looked to the mysteries of nature first and foremost, because they were able to observe the wisdom of the cosmos in nearly all earthly functions. By seeking union with this cosmic energy, once called Sophia, now called Earth or Gaia, the individual could receive inner wisdom and further instruction in their own spiritual cultivation. Along these lines, we're able to understand initiation as something that did not originate for the purpose of gaining access into any particular society or order. It's true purpose was for the discovery of one's own inner path. Thenceforth, when that candle was lit, the spirit of the anthropos would be illuminated. In its revealing, the true magnificence of not only our Magna Mater - Sophia, but our ancestral roots in the pleroma or divine hub of the galaxy would be realized. Furthermore, we would know that our roots were universally connected to the neural network that exists between all the cosmos.

Initiation by our new and rather retro conceptual ideology of furthering one's own inner wisdom, derives from an individual challenging their preconceived notions of the greater cosmos. They do this, not by challenging their concepts, which revolve almost completely around faith, but the beliefs they've developed through those concepts. And as stated before, our faith is merely a trust or openness we have for the greater Universe. Our beliefs on the other hand translate to the things that we Love or value, which are quite simply opinions. So by challenging our own personal beliefs, we allow ourselves the opportunity to expand our knowledge of the Universe, and learn to be less firm in our devotion to belief systems. If a person is unwilling to challenge themselves, then they're submitting to ideals that are basically programmed into their mindset starting around the moment they began verbal communication with the world. In order to initiate oneself into a higher wisdom, and to truly explore the divine picture, a person must first challenge the pre-existing beliefs they've developed about the purpose of life, which begins by observing the nature of our earthly environment.

This art was learned through what are called "The Mysteries," and was part of what seems to have been a circuit of Universities stretching from northern Europe all the way to the far east of Asia. For example, why are there stories of Jesus and Dionysus across Europe, North and Central Africa, The Middle East, and Asia? Could they have been gifted students of these schools, or perhaps... exceptional teachers with profound insight? Furthermore, I would imagine that this circuit(largely secretive) now encompasses the entire globe! I mean why wouldn't it? Every culture has a tremendous amount of insight to offer, especially when it comes to correlating myths, theory, science, and philosophy. By this interpretation we can deduct that no one school is the end result. In order to know... you must know and then decipher how you will continue measuring what you don't know. Basically the mysteries circuit allowed a person to decide which path they'd go on(light, dim or dark), and not be forced down or expected to accept any particular school of thought as gospel - like what we find in religions today. The initiate, or student, was allowed the ability to know certain truths and cosmological perspectives through extensive research and ritual, and in turn, be offered higher wisdom - allowing them to formulate his or her own opinion. From this perspective, it makes a person wonder if religions are actually the result of certain students graduating thesis', or if they're simply the result of one school of thought seeking dominance? I'll leave this question for the reader to decide...

Hermetic Philosophy - The 7 Keys to Universal Thinking

The Principles you are about to receive, which are referred to as the 'Hermetic Principles," many may be familiar with. They're the key elements of a profound philosophy called "Hermeticism," which is claimed to be the oldest on the planet. In fact, some say this philosophy predates our earliest civilizations some 6,000 years ago, and is of a previous global civilization! However, the principles brought forth by this philosophy have hardly ever been at the forefront of our mainstream attentions or ideologies, which is what makes them particularly interesting. What's especially odd, is the recognition of the finely crafted hermetic thread that's interwoven into our cultural and scientific fabrics. As a matter of fact, once a person becomes aware of these principles, it's nearly impossible not to witness them as a measure of every part of our cultural and scientific ideologies. Furthermore, these principles, though allusive with their cryptic complexity, when broken down, hold explanation to nearly everything that can be observed in not just the world, but the Universe itself. In light of this statement, I will firmly state that these principles are of extreme importance for our understanding of the elementary nature of not just the anthropos, but the greater cosmos.

Interestingly, but to no surprise, all religious and philosophical traditions, in one way or another, refer to these principles or use their expressions to convey their messages - whether they choose to admit it or not. The parable in fact is a similar form of speech to that of the hermetic philosopher, because of its cryptic nature, which was purposely engineered as an oral tradition passed on as they say, "from lip to ear." So when Jesus said, "Whoever has ears to hear, let them hear," he was announcing his intention to reveal hermetic wisdom. Therefore only those who were initiated into this form of speech could know the true intention of his words. This of course is the real reason why the pharisees had him killed. He was revealing the secrets of what are now called "the mysteries," which were forbidden to be public knowledge. The mysteries were only allowed to be revealed through the esoteric tradition of initiation. And without a doubt, just like it is today, the right of passage for initiation was only granted to the elite, religious classes, and those who exhibited exceptional ability(in myth these are star children - ie. Jesus, Dionysus, Perseus, Krishna, Moses, Akhenaten, Osiris, Hercules, etc). The reason for this is quite simple - direct knowledge of these principles offered a way for the mind and spirit to awaken their inherent powers, which is why they were forbidden knowledge - especially for the commoner! This is why they were publicly looked down upon as heresy, occultic, or even alchemical hub bub, because they posed a direct challenge to the religious and political authorities.

    In an overtly aggressive approach towards its real intention, the mysteries have been demonized by political and religious institutions dating as far back as ancient Greece and Egypt, mostly by monotheistic societies(groups claiming one doctrine or school of thought to be the only truth). The knowledge of these philosophical principles were basically forced underground and became a full on part of a semi-secret circuit of universities stretching across much of the ancient world. This brings rise to the concept of "Mystery Schools," which were institutions that investigated the mysteries of life. For some time, after Catholic-Christianity sought to rid them from the world(persecution of the Telestai, Gnostic or Pagan religions) after the council of Nicea - they lay dormant. However, interest in the mysteries were resurrected by the Knights Templar during the first crusade(did they manipulate the Vatican into the Crusades in order to uncover undeniable secrets which were a key part of the mysteries circuit?). After their demise in 1307 by papal mandate(when the Vatican got hip to their game), the wisdom went further underground, passing into the realm of secret societies. Where it went from there is left to much debate, because it became high profile religious treason to practice any of the ancient teachings in the open(even though political and religious elite continued their mysteries education). This brought rise to the period of the Inquisition, which for truth's sake was to completely stomp out the ancient wisdom of the mysteries, allowing the Catholic Church to dominate all matters of knowledge and spiritual cultivation. And if you think the inquisition was bound only to Europe, think again, because Catholic-Christianity nearly obliterated much of the civilizations in the new world, as well as anywhere else they went! However, as we all know, the truth is something that can never die, because it exists at a level which is far deeper than our physical understanding.

For the person seeking to "know thyself" in this wonderful experience we call life, these principles, which are easily observable in Nature, represent the elementary wisdom brought forth by the Magna Mater of our anthropos, or, as she has been named - "Sophia." If there's anything in your life you wish to try and figure out or understand, whether it involves science, philosophy, or biological relationships at the micro and macro scales, these principles will assist you greatly. If you're looking to find correlation between your life and the outside world, whether it be the archetypal relationships of self vs. other or self vs. environment, these principles will assist you greatly. The applications are truly endless!

Without a doubt, these principles exist as a puzzle piece to our birthright, and a gift given to the measure of the divine experiment, now unanimously referred to as "the human being."

You see, not only are the hermetic principles thought provoking, but they're referencing a specific order of nature that can be witnessed in all things. Furthermore, they seem to be the alphabet or bridge between our polarities, as well as gaps in thought. For example, with use of these principles, a person can correlate the relationships between science and religion, as well as their comparison to astronomy, and conduct their analysis in a universal manner. Presumptuously, we can deduct from this explanation, the idea that one's mastery of these principles has the potential to unlock a greater understanding of our cosmology and all of it's processes - Which is why open debate or exploration of these principles is considered esoteric and occultic. After all, what purpose does it serve knowing these principles? Who do you think you are... God? This of course is a poor attempt at humor, but nonetheless seems to be the main argument as well as ignorance towards the matter today(wouldn't it be nice if we could all share this education, rather than the bogus educational systems given to the majority of us?).

In light of this debate, the Hermetic principles are considered the fundamental keys on the philosophers keyboard - the 7 basic notes. They are the 7 fundamental principles found in most all sciences. Hence these 7 principles, claimed to be Hermetic or of Hermes Trismegistus(thrice great), are what I refer to as the "Universal Principles." Understand them, but even more - respect them, because they have the potential to unlock the patterns of Nature that were once thought mysterious - revealing amazing truths about our great cosmology.

The Principle of Mentalism - "All is Mind"

If you are made, or created in the One image, you are a "Creator in the flesh," which means that your mind creates your reality. In this principle of Mentalism - "All is Mind." Everything in the entire Universe is a product of the One Mind referred to as the ALL, The Creator, The Universe, The Ether - whichever you choose to call it. This One

Mind - this ALL, is what we also refer to as "Consciousness." And considering that Creation is the fundamental purpose of existence, it means that Mind is what creates IT. The Universe in one way, shape, or form, was created from one single thought, question, contemplation, or expression, and like a rock dropped in water - has continued to ripple or emanate outward. By another interpretation, we can say that, one thought has manifested itself into an infinite amount of thoughts - venturing further away from the source point or original contemplation of the mind that created it. Therefore, we are given the biblical translation in Genesis, "In the beginning there was the word." And a word is obviously a thought or vibration that originates from Mind. However, when we observe the structure of the Universe, we can confidently say that thought is both creatively conscious and subconscious. It's conscious to the point that it's always focussed on creating, and subconscious because it continues to give to everything that it's created without having to think about doing so. Why else would a human being be both conscious and subconscious if we were not created in that One image?

Mentalism requires us to reconsider all things. For instance, the light from the Sun is not just sunlight, it's star light. And star light is not just star light, it's THE Light. And The Light is consciousness, or what is referred to as "The Light of Mind." Afterall, who is the one deciding this fact? Who's observing it? It's your eyes, your mind, which observes this phenomenon. So the source of all light is actually coming from your mind! After all, light wouldn't exist if you didn't have occipital lobes(source of vision) to distinguish light from dark. Thus the Universe is, because it is of Mind.

All is Mind, because Mind is the creative force of the Universe. A human being can visually imagine something and in turn make it a physical reality. A wheel was never a wheel until a Mind created it through thought, and then put into form. The Question remains though, "are your thoughts really your own?" Or, "are they a part of the conscious pool of thoughts streaming from the One Image?" To answer this, we have to understand that a thought is no different than a wave on the ocean - one thought is merely the representation of the greater mind. This does not rule out the possibility of originality or uniqueness, but merely the notion that thought originates from the highest of the high. All is mind, and you are the representation of that mind, in the flesh. It doesn't mean you are supreme or perfect, it just means that you are. Even more, you are The Light of Mind existing now - wishing to know itself in this particular form - a human being.

The Principle of Correspondence - "As above, so below."

All things exist in corresponding nature with the One Image. Now when I refer to the One Image, I mean Creation itself - the Mind that makes all this possible. So what we experience in our own lives, is a representation of what the Universe is experiencing now. To further, the human being can be considered a micro-universe. It has close to a billion trillion cells, all existing separately from one another, but pulled together by an energy that coordinates motion and function. A human being cannot consciously focus on each and every individual cell in its body, otherwise we would spend all of our time trying not to fall apart. Thus we have a subconscious or unconscious mind which handles these processes. If a cell needs something, it sends an electromagnetic frequency to the brain, which unconsciously responds with an electrochemical response, as if to say "here you go my Love." The Universe continues to consciously create, but also subconsciously continues to provide for what it has created. The Human being is said to be made in the One Image. That means that the processes of the Universe and the Human being(a micro-universe) are in corresponding nature. And so the saying goes, "As above, so below." "As it is in heaven, so shall it be on earth." This of course relates perfectly with how creative this planet truly is - it too is acting in corresponding nature with Heaven above. It's all about one continuous process of creativity - operating by the same relative principles, which are corresponding with the One Image.

The Principle of Vibration - "Everything is in motion; nothing is at rest."

Like the principles of a wave, everything that is, exists in frequency; a vibration. Not one thing that exists is still, even though some things give the appearance of being still. A rock, though appearing still since the beginning of all things, at it's atomic level continues to vibrate. Light appears to be constantly bright, but when slowed down shows that it exists in between dark pulses. Therefore, when we observe a lightbulb from a distance, it flickers or sparkles, because of it's pulsing nature. Yet when we view the light from up close it appears to be motionless. And so we can see that stillness or form is an illusion of frequency, which can be altered by observing such frequencies from different distances of time or space. By this measure, all things that are, are of frequency. The only difference however between light and a rock is their level of frequency. Light exists at a very high rate of frequency or vibration, where as a rock, which could be considered the grossest form of matter, exists at a very low vibration. Either way, both are in motion, because they're vibrating.

The Universe, 99 + percent of which is energy, exists at a very high rate of vibration and a very low rate of vibration. A human being, at least the physical aspect of this being, is at a lower rate of vibration - closer towards the middle, and thus, can only see a certain range of motion. We refer to this as our window into the spectrum of light. What we see is only a sliver at the very center of that spectrum. To our left are the illuminating frequencies of ultraviolet light, and to our right are the radiating frequencies of far infrared. But nonetheless, it doesn't matter which direction we observe, whether it be left, right, or center - all is in motion or vibration.

Everything moves; nothing is at rest. Stillness is but an illusion of higher and lower frequencies existing outside our being's ability to observe the total spectrum that is Mind. So know that everything you see; everything that is, is vibrating and emitting a frequency.

The Principle of Polarity - "Everything is in opposition with something."

Everything is consistent with being in a polarity with something else. For instance, Love is in opposition with fear, light with dark, positive with negative, good with bad, body with spirit, and so on. That's the nature of life, in that, in order for one thing to exist, it must be opposed by something else. High frequency is in opposition with low frequency. A wave has its crest and it's trough. Are they separated? Not at all! They're merely at different degrees of the same one thing. Fear is in full connection with Love. Does that mean that Fear has a power over or can be greater than Love? No, but you can't have one without the other. If there were no Fear, there would be no Love. If there was no dark, there would be no light. Everything is in opposition, but only for the purpose of experiencing the full nature of every one thing. A human being experiences birth, and they also experience death. However, what do we get when we add birth and death together? We get Life! The same goes for everything else. Everything must exist in opposing nature in order for the One Image or Great Mind to fully know itself. And because of this, we have to know that every one thing that's in a polarity with something else, is merely expressing the polarity of one total thing.

The Principle of Rhythm - "Everything sways like a pendulum."

Since everything is in a state of motion, as well as in polarity, every one thing must sway in each opposing direction. Waves rise, and then they fall. The ocean has an ebb and a flow. Day becomes night, and night becomes day. The lungs inhale, and then exhale. Everything has a rhythm! Some may be more extreme than others, but nonetheless it's the way. In order to like something, you must equally experience the feeling of disliking something on the opposite side of the pole or spectrum. And since life is in motion, you must sway between both poles. Does that mean that you have to constantly experience a manic nature? No, but it does mean that if you apply yourself to a polarity, you can expect to sway in out of the frequency and rhythm of that polarity. If you seek to live only at a certain level of good, you must equally experience as much bad. It doesn't mean that you have to respond to the bad with bad! It just means that you are swaying with the pendulum of life. Everything in motion, sways rhythmically in a polarity. That's why we're told to "find our rhythm."

Consequently, because everything is in motion, and every one thing is in a polarity with something else, then we must experience both aspects or frequencies of that one thing. Therefore like a pendulum, we sway back and forth between the two poles of any given thing - ie. waking and sleeping, standing up and sitting down, the blinking of the eyes, etc.

The Principle of Cause and Effect - "Every Cause has its effect; every effect has its cause."

Cause and effect is one of the most well-known universal principles. For every action, there is an equal reaction. One cause equals one effect, and vice versa. Most only think of cause and effect in terms of action and reaction, however, this principle goes beyond just our physical expressions. We're also energetically, and even more, mentally creating actions all the time. One thought does not stop at one thought, but continues onward to produce more thoughts. So if we have a positive thought, it will produce a positive emotion. On the reverse, if we have a negative thought, it will follow with a negative emotion. Thoughts and feelings coincide, because the rational and emotional brains are in-sync. However, each are still the result of the other. The Heart initiates the creation of the physical body, and the Mind settles into the whole thing - only to exist in a symbiotic relationship with the heart.

Moving further into this principle, we see it correlate quite nicely with our previous principles. Everything is a product of mind, which produces vibrations that are in polarity, which also means everything must be in rhythm to experience both aspects of every one thing, and every cause of motion or sway of the pendulum creates an effect that perpetuates the cycle. Starting to see the larger picture? Each principle will always be in a direct relationship with the others, because life is a spectrum or flow of many different variables existing at one time. When you begin to observe life in terms of the principles of Universal motion, you begin to see that it's all interrelated and One. This is the essence of Universal thinking.

The Principle of Gender - "Everything is a combination of both male and female."

Every one thing, though usually declared one or the other, is a combination of both male and female. However, the determination of what sex one thing is, is only a perception of frequency, or where vibration exists on the male-female polarity. Meaning, everything that is the One Image is inherently both male and female. Be that as it may, it's also an expression of what frequency the One Image is deciding to exist as. So when we see a man or woman, we know that Creation has formed itself on one side or the other. However, both male and female still have an element of the other within them. So if you've ever heard someone say, "You must get in touch with your Feminine/Masculine inside," you know where it originates. It isn't just a clever way of saying, "be more centered in your way of being." It's a realization of the One Image's androgyny, or, polarity existing within you. You're being, is how Creation chose to express itself now. Therefore, it's always a matter of Creation, and how Consciousness decided to formulate itself in you. We're given a tremendous gift with this experience, and the One behind us is "Everything that Is." So it matters not whether you're male or female, because you're still It!

The Hermetic Principles are golden, and undoubtedly the philosopher's guide to being more Universal in thought. If you look back through our history, you'll see every great mind, both male and female, speaking the truth about their time in a familiar nature, tone, or rhythm. With respect, we can understand hermetic philosophy as more than just principles, but the Mind's alphabet for figuring out the true nature of not just humanity, but the Universe itself!

Use these principles wisely, and never neglect their inherent purpose - to provide you with the essential tools to be the focal point of consciousness - made in the One Image... of you!

Chapter 4

Life's Eternal Wave Motion

In order to understand the nature of the human game, we all need reference points. These points are usually found as indicators in the background of our conscious attention, and are awakening principles to what many may call "the path to Universality." However, the issue that we typically face, and the main reason why we aren't consciously aware of universal principles, is actually a product of human advancement. Our evolution of thought in many ways has hindered our ability to take in the entire image! You see, thousands of years of working to increase our focussed awareness, has resulted in the development of shorter attention spans. Therefore, we live in a world existing of what many feel is a conglomeration of separate aspects existing together. The price we pay for this specialized focus is an alertness to smaller and smaller details, where we can hone in on one object, but at the price of ignoring the background. As a deduction, separate subjects and events are not taken seriously, which makes us live almost completely for definition and form. As a result, our focus typically exists only within the illusion, or, what we've developed with a more centralized or central vision - perpetuating the inception principle, where the dream becomes the reality. By this idea, we continue to formulate more and more beliefs that give authority to our already separate physical environments. This gives us the psychological interpretation of the "rorschach blot," where we only take in what we can make out, or, what our thoughts focus on. By increasing focus, we've generated a reality that's cut off from our peripheral vision, which is what allows us to take in the entire picture. In fact, our world today appears to be a system of inseparable differences. To some extent, this insistence of further sovereignty for the finite details, has made us ignorant to the existence of our underlying or innate functions. Wherefore, when we pull back and take in both peripheral and central vision at the same time, we're able to observe a different reality. By this measure, we're now observing our environment as opposed to trying to figure it out. Here's where we're able to find the most clear and definitive truth we're constantly exposed to... Life is not separate at all! It's actually an interconnected web strung together by wave motions!

We find waves everywhere we look - whether it be on the micro or macro scales(principle of vibration). After all, waves are synomatic to frequency and vibration, which is the result of universal continuity. Thereupon, to truly understand this wave nature, it's pertinent for us to examine the most obvious wave motions on this planet, which brings us to one of our favorite peripheral observations... the ocean.

When we observe the ocean, not only are we using both our central and peripheral visions simultaneously, but we're able to observe one of life's key phenomenons. With wave after wave rising and falling continuously, as if time pulsed with it, we're able to identify the illusion of separation. Furthermore, we can predicate that waves interconnect to form one larger function. Both our central and peripheral visions, therefore give us opposite ways of how to perceive life. However, they're not separate aspects, but merely a degree of how open we are to our environment - Are we narrow or open-minded? So we can extract from this idea, that polarity plays into how we're perceiving our experiences. Correspondingly, we can assert that the ocean too falls into this perception based game!

The hypnotic operation of ebb and flow keeps the waters of the world in tidal resonance. In this state, the ocean remains in motion. Even on calm days, when the water sparkles in still reflection, and all the world seems to reflect back at us - the ocean is still in frequency. However, it's merely expressing itself in larger and less pronounced intervals. So it doesn't matter whether the seas are stormy or calm, the ocean is always in euphony or melodic resonance - the same as everything else!

You see, oceanic observation affirms that all life can be broken down into wave motions. In addition, our unified view of centricity and periphery shows us that it's all one rising and falling process of corresponding patterns in nature. From this conclusion, what the observer eventually recognizes - one single wave is only a smaller representation of the larger ocean. This one wave can be described as the entire ocean existing within a smaller version of itself!

So on the one hand, a wave is experiencing a central or individual journey, but on the other it's expressing a single thread in a larger peripheral fabric. Does this mean the individual wave is unimportant? Not at all, because Creation doesn't work that way! Everything as it is in it's present state of uniqueness, is the One Image existing now. So when we focus intently on that one wave, we're actually observing the spirit of everything moving within itself - the finite therefore expresses the nature of the infinite, and the infinite expresses the finite.

This wave nature doesn't just exist as we typically think it does when it comes to the ocean - it exists everywhere; in all things. Light for example operates in wave intervals, because light is pulsation. So between every pulse of light or brightness, there's a pulse of darkness. Light rises to be seen as something bright, and falls into what is perceived as dark. However, the wave remains continuous in it's own nature, so that light continuously pulses to it's own beat or interval. The human eye naturally cannot see the wave nature of light from up close, but if we slow down the frequency of light from let's say - a lightbulb, we can see the appearance of a flicker. This is because waves have to rise and fall in order to maintain momentum. Thus light and dark play off each other. The result is one total expression realized through opposing forces, which are crests and troughs. The question is though, "how quickly is the wave moving?" Light works in very fast intervals, giving the appearance that it's always bright. Is there a wave that takes more time in it's rising and falling? Well, look no further to where you sit, lay or stand... A human being is also a wave! After all, we're born, we grow to a certain point, and then our wave falls downward towards death. And it's from this wave nature that we see an undeniable truth in everything that exists, which forces us to question, "is death really the end?" The answer of course is no! Death is just a human beings epitomic version of Change, which is what the Universe does - it changes!

All things are a substance of energy that rise and fall continuously. The energy that is, never dies, but merely transforms or transfers into another expression of itself. The question is, "in what direction and with what?" Do these waves merge with other waves? Do frequencies change the rate of their pulsation to take new shape? Well, of course they do! By the nature of creativity, Change is a universal constant, because nothing remains the same. Everything must experience its own involution and evolution in a becoming and unbecoming process. Thus we're all rising and falling in our own individual wave, which is continuously evolving with the rest of the Universe - into something else! This is how and why human beings have gathered the notion that a person can raise their own frequency... If we're all moving in a direction with the larger wave, or even a point pulsing outward(like a dropped rock creating ripples in water), then we're constantly in motion. However, the question remains for this three dimensional mind - into which direction will we go? This of course gives us our heavens and hells. But let's step away from that for a while, because life should not be about the destination, but rather how we're experiencing our own wave, right now, in this very moment.

Are you fighting who you are and the nature of how things flow? Can you, really? Well, in some ways you can, but in the grand scheme, we're all going somewhere. Both our periphery and central focus play off each other to show us life's need for opposing nature. Opposites therefore create momentum, which perpetuates our continuation as an energy in this Universe. In fact, opposing natures are what sustain universal motion. Thence, opposing forces need be observed as inseparable points or poles on the end of one larger scale. The same as a wave has its crest and troughs - both crest and trough are not separate from each other, but make up one rising and falling wave. When we recognize this, we're able to determine that our changing patterns and experiences, as well as emotions, are a direct result of life's need to always express itself with the other. Forces of nature continue changing their rate of pulsation, and even direction, for this very reason.

Life should not be a fight when looked at in this regard. It's only a fight if we're seeking to hold back the nature of ourselves. Thereupon we all must realize that waves are meant to flow freely with the larger ocean, which continues to change. By this measure, waves have the opportunity to know the larger ocean as well as their own nature existing within that ocean.

Think of life in terms of a wave process. Know that each and every one thing - no matter its composition, is in a process of becoming and unbecoming itself; rising and falling; expanding and contracting it's own energy. Just like the Creator of All That Is will inhale and exhale a thought into the Universe - you are an evolving thought of this Creator, which makes you the Creator in the deepest form of Itself. You're a Wave - a frequency of all that is, and eventually you'll make your way back to yourself. In that moment, you are the entire ocean, moving, and creating more waves - in rhythm with your own nature; the periphery and focus of the entire environment...

Like an Element Flows

When one thinks of philosophy, they immediately seek to find some form of truth that'll convey the meaning of life. For the pragmatic mind to understand Nature, which in the beginning doesn't seem to make sense in the slightest, we first look to its fundamentals. The elements of Earth, Fire, Wind, and Water become philosophies of their own when observed carefully. Yet for the sake of using these philosophies only as an example, for now, we'll revert back to the element which reminds us of a wave... Water.

"Empty your mind - be formless, shapeless. Like water! Now, you put water in a cup - it becomes the cup. If you put water in a teapot - it becomes the teapot. Now, water can flow, or creep, or drip, or crash! Be water, my friend."
- Bruce Lee

Water is a conundrum, a comfort, and a continuous measure of the Universe. When we think of water, we think of what? Well, we typically we think of hydration, but more importantly - we think of the flow of Nature. Water represents the largest portion of the surface of the Earth, and quite literally dominates her landscape. Yet water has a rhythm that relates to more than just the dynamic of the Earth. In fact, it relates quite well with the rhythms of our bodies. After all, a human being is largely a concentric flow of water molecules - moving us beautifully in a largely unconscious way. So just like when we determined that waves flow synonymously with the larger ocean, as does a human being with the waters of their own life.

Water is one of the fundamental essences of life. Without water, which we're between 40-90 percent depending on our age, life would be of a lower vibrating field - similar to the frequency of a rock. Water keeps our bodies chemical processes functioning at a high level, even when we're not operating in a desired rhythm.

You see, we're not only electromagnetic beings, but electrochemical. Water is the substance that makes this possible, because it lubricates every aspect of our being, in order to allow electrical information to pass through us efficiently. In a way it makes us the perfect conduits for electricity, which is none other than plasma photonic light, or electrons in motion. So when the supposed gravity of our world seeks to compress or sink us, and electromagnetic forces push and pull us, water stabilizes the effects and keeps us buoyant.

To understand this element; this philosophy - a bit more, we must observe it in it's unimpeded state. So if you've ever watched a stream flow down a mountain, you know that nothing will prevent water from reaching it's destination(wherever that may be). It is said that, "all water eventually reaches the Ocean." So from the highest peak, water will flow thousands and thousands of miles to reach its root home. Even more, water will not be held up, even when it pools and appears still. Eventually, water always finds a way, whether it be through evolution, erosion, or evaporation.

Going back to the stream, when water reaches a blockade, or begins flowing to create another channel, it sometimes gives the appearance of being held up. However, if you continue to watch, eventually water finds its way around the obstacle and continues it's journey.

We are all molecules of water flowing down the mountain, seeking to reach our root source - the ocean(The Truth). If and when we feel held up, if we let go and move with the flow(Life), we'll eventually find a way. This philosophy is the key to understanding life's motions.

If your being can be as fluid as the element of water, you'll be unimpeded in your flow, and in time become one with the larger Image of yourself. Likewise, if your mind can exist fluidly in a field of conscious thoughts, you can visualize and or imagine any possibility possible.

Water is a source and a necessity of living, but even more, it's a philosophy of being - a way of motion that represents the nature of the One Image. So in keeping with the philosophy of Mr. Lee, "Be like water, my friend."

Chapter 5

The True Nature of Change - "It is"

"Change, and everything is change(nothing can be held on to), to the degree you go with a stream, you see, you're still - you're flowing with it. But to the degree that you resist the stream, then you notice, that the current is rushing past you and fighting you. So, swim with it! Go with it! And you're there - you're at rest."
- Alan Watts

Change, stated by every philosopher to date, is the one constant that is, well, unchangeable! But we human beings certainly like to think otherwise. As a matter of fact, we've developed a civilization that's fundamentally resistant to change. Why is this? Many attribute this flaw to be a direct result of material mindedness. Thus, if one lives solely for the material world, then everything they recognize as a part of their lives is at the mercy of the great wheel in the sky. This material world fades moment to moment, as if time exists to strip apart all creation and return it to dust. As a consequence, the material minded person lives solely for holding onto life's objects, and the things they feel they've gained. They hold onto their homes, their possessions, even their relationships, because they realize time will eventually take everything away. Yet this idea proposes an even deeper concept - one that exists on the far end of our anxiety, fear, and worry. If we look into it far enough, we find that a material minded beings main objective is learning how to defy death - and in many cases, trying to overcome it! Why else do we try and turn our lives into a set of balanced routines? Is it not because we're trying to control the inevitable? Our balanced routines are attempts at creating order in a constantly changing environment. As a result, we end up creating the things we fight for incessantly. Well, why shouldn't we? It's our lives, and these things belong to us, right? Yet, here's the contradiction - one day we're going to die, which means that life and all its possessions are on loan. Therefore, we can't really own anything! Which must mean, that somewhere outside our central vision - somewhere in the periphery, is the truth - life is a temporary experience.

For example, a fruit tree grows from seed, tall and wide - producing for countless generations. But, eventually it submits to life's need to change. And so it returns itself to the Earth, where it's broken down - fertilizing the future environment.

We have to understand that everything around us is changing - the whole Universe. When we choose to remain unaware of it, then we're living at the mercy of nature(change). And that's what the majority of us continue to do. Somehow, we resist the truth, and defy, defy, defy - playing completely into our own delusions. The idea of losing ground anywhere - in anything, becomes a persistent background fear that we perpetually fight - creating the chaotic daily lives we're all seeking to manage.

> "The worldly hope men set their hearts upon turns ashes - or it prospers; and anon, like snow upon the desert's dusty face, lighting a little hour or two - is gone."
> - Omar Khayyam

So rather than us learning how to live and live well, we fight the current that can't be resisted. We have to realize, there's no such thing as a wave that doesn't rise, nor is there a wave that doesn't fall. Every one thing that's created is a part of the one process of becoming and unbecoming itself. It is the Tao, or the Way - the very nature of the cosmos.

When we resist change, we're not only resisting the flow of the greater Universe, but also the flow of our truest selves. When we look hard, we see that there's really no other problem for a human being - We're constantly fighting a changing nature!

There's two sides to a human being, the same as there are two sides to a wave. Therefore, human life is a unified duality, or, an expression of the two natures of one thing. Hence we're the light or life principle, and the dark or changing principle. Notice how I'm not using the dark principle as death? I'm doing this because I want you to know, the death that many of us believe to be everlasting non-existence, or something that we should fear as an end to our energy, is not in the slightest bit accurate! We fear death, because we fear change. And because we spend most of our lives trying to figure out how to control this experience, we worry that we'll come up empty and not meet our expectations - we'll not reach that pot of gold at the end of the rainbow. Well, just like going to sleep at night and hoping to have good dreams, we fear what will happen on the other side will be a nightmare. It's all relative! Why else do you think Christianity tags us with a final judgement? Is it not because we're always judging ourselves and hoping things will turn out alright in the end? We wish it not to end in darkness, but in light, right? Yet when we look deeper into the physical nature, we find the darkness that we perceive as darkness, to be merely an illusion created by our own physical eyesight. We see light in terms of a small fraction of the spectrum that is light. So we would have to know, when that waterfall comes, we're not falling into an abyss, but into a deeper state of being. Therefore, the ability to let go is the only real control we have in life. As a correlation, a flower comes into itself and blossoms fully and beautifully, but eventually lets itself go. It doesn't miss itself when it's energy moves on. It simply lives every moment of its physical life presently, and when it's time to move on, it moves on well. In the same way, a human being should always be prepared to die well. And we do that by living gratefully now.

If you can see that resisting change is what's holding you back from living fully now, you can put to rest your fear of dying tomorrow. You can exhale the desire to control this moment, and let yourself rest. Then you're there - in unity with everything that is; free to appreciate life and all its wonder. In fact, when we stop resisting change, our perceptions alter. Thereupon, we're able to understand the changing world as something that's no different than what we refer to as the "heavenly," because we've completely let go of form - we've exhaled. This exhale, breathing out or letting go, has been described in Buddhism as the ultimate experience of Nirvana. In this state, a person has embraced Life's changing nature and become one with the finite and infinite. They've let go of trying to hold form, thought, or any other idea that life will remain the same - and they breath out. Our salvation, deliverance, and/or liberation is the result of the this very idea. It's the realization that we can't hang on to ourselves; this being or body - nor do we have to try and hold on to anything, because it cannot be done! And so freedom can be described as the experience that's achieved when we let go of defending ourselves - this form who is afraid of changing. By the same measure, in Christianity, a person must let go and or surrender themselves in order to allow the current to bring them into "salvation." Respectively, we find salvation, Nirvana, liberation, and freedom are all expression of the same inherent principle... Only when we stop defending what's impossible to defend, are we capable of experiencing the Universe fully in this One moment, which is the only moment that has and will ever exist... Now! Life is eternally here and now, and is always changing! So it's time for us to get with it - flow with it, because it's who we are and what this is all about!

Getting Out of Your Own Way

In order to understand this system of waves, we have to know what it means to get out of our own way. You see, we're all searching for that one very important thing - God, Allah, Nirvana, The Creator, Brahman, Enlightenment, etc. We all wish to feel as if we're fully connected to this One power that creates and therefore sustains this Universe's emanation. Yet what we forget to realize, is that for every one thing we learn, we develop one more illusion that keeps us from feeling whole. So for instance, by me knowing that I'm just one small speck of focused matter and energy, existing on a planet with billions of organisms, revolving around a star in a solar system with 8 to possibly 10 planets depending on whose view, in the outer part of one of four spiral arms or density waves that make up a rather small galaxy with roughly 300 billion solar systems, which is part of a Universe with an estimated 2-400+ billion galaxies - I've now carelessly expanded my desired field of connection to a near impossible degree of comprehension. That's quite a large field to connect with, isn't it? So you see, by trying to connect to this massive organism we call "the Universe" or "the one line or verse," I'm running the risk of cutting loose or even grossly neglecting the connections I have here around me(family, friends, the Earth, etc.). And by developing the alibi which nearly everyone on a spiritual path comes into contact with, "everyone is asleep," I'm further isolating myself from the Now that exists right in front of me. For this reason, by trying to expand my consciousness, I've separated myself further from being here and now. This is how the truth seeker continuously gets in their own way.

Eventually, we all have to decide where we wish to be. Do we seek to be in the "Now," which exists solely in our field of "here," or do we wish to go all the way? Well, there are two things wrong with this decision. For one, why do we have to make it? And two, why are we separating the "here" connections with the "out there" connections; the central with the periphery? You see, if you wish to be connected to the full nature of this Universe, you must let go of all separations in the here and now. In fact, when this Universe is broken down, we find that it's all based upon descriptions of just one experience - one faith or "openness." It's been referred to by the scientific community as quantum entanglement, or the notion that subatomic matter has never lost it's inherent connection to the original nature of the Universe before it's emanation or inception. Which means that all matter, subatomic to galactic, is still fully connected. Most religions quite beautifully state that everything is created within the "Spirit of God," which is basically saying that everything exists within one experience referred to as God or the Universe. Even you, the one who perceives to be separate from all of this, are fully connected to the far corners of the Universe. By this definition, the Great Mind of the cosmos is as much you as you are it! After all, it's you who is doing the measuring, are you not? It's your mind that's calling the shots - you're deciding what is and what isn't!

So when any decision is made, let go of the connectivity question, and make the decision based on your own measure - "What kind of Universe do I live in now?" or "What kind of world do I live in now?" Therefore you make decisions based on your own wishes. "I want this to be a Universe which celebrates life." Which is the same as saying, "I celebrate Life!" "I want this to be a world that Loves." Which is the same as saying, "I Love!" And so it is, it is so! Because you're the one doing the measuring, it means you're not only connected to the entire Universe, but you're in fact exactly what the Universe is doing right now. And furthermore it's doing it through you!

So, how will you measure yourself in this moment? Which is the same as asking, "What kind of Universe do I live in now?"

Chapter 5

The Art of Discipline

Hopefully you're beginning to feel liberated and perhaps awakened to what this Inward Way is all about. Going Inward is about letting go of control or holding onto life and it's experiences. It can also be described as "not resisting the nature of yourself." It's quite simple when you observe it plainly, but as we all know, the simpler something is - the harder it is to put into practice.

As conscious beings, we have an imagination that's truly limitless. However, because we live in a belief-driven world, we're subjected to belief structures that are mainly subconscious. And because subconscious means something is below our realm of conscious attention, we're typically unaware of most of our beliefs - particularly the beliefs designed for the purpose of control. These controlling beliefs, of course, are what hold us back and limit our potential. They mainly represent the deepest beliefs, which hit us the hardest - when we least expect it. Experiencing such beliefs, is similar to when a dog is held up by its leash after taking off in a full sprint - it comes as a complete shock, which can be very painful. It's the same as a student in grammar school who has a great idea they wish to express, and the teacher tells them, "keep quiet!" Of course the child feels completely shut down. These deep beliefs work in roughly the same way, because they're designed to keep us protected and within a known and stable rhythm. Yet, as many of us know, often times these beliefs do more damage than good, because they keep us thinking and feeling within a "norm." Typically, these control mechanisms coincide with the illusions set in place to keep us from seeing the overall big-picture illusion, which in the end is quite simple... "Life is truly a product of our own choices," or "a matter of our own free-will."

Ultimately, we must come to terms with this truth. And furthermore, understand that we create all of our own authorities, and can thus change them anytime we like. For example, the student is really in charge of their own life, and the dog can bite through the leash if it chooses to. Now of course, students do need guidance, as well as certain dogs. The point of the matter is, you - the human being who seeks to measure all things, are in charge of your own life! And anyone who tells you otherwise has bought into the fear of change, which plagues the belief-driven world - we're typically not allowed to feel unique, or, express ourselves fully as individuals.

As we begin to let go of our illusions, the pains that arise from fighting change begin to disappear - they fade away with the current. But the question is, "how do we continue to remind ourselves to let go? Isn't that the nature of life - forgetting that we're not in control?"

You see, life is an experience of many moments of divine revelation, or times when we say, "This is it, I've got it!" And right when we settle into that knowing, we realise, "What have I actually got?" So as we try to grab hold of it tightly, it slips away from us, because it was never really ours to have - It was just part of the passing current. This is the nature or wave of life, which we move through in periods of knowing and not knowing. It can be referenced as one of life's built in mechanisms that teaches us to let go, and move WITH nature. After all, we're not here to learn how to attain the Creator's wisdom, because we are that wisdom! How can you attain yourself if you're already you?

In Zen practice, often times the master or guru will physically strike or shout at their student as a form of shock treatment. This jolt is meant to wake the student up. It's similar to when we find ourselves falling within a dream and the kick of the fall wakes us up. These not so subtle reminders only bring us back to where we are already. They help remind us that we, like waves, are always in motion with the larger body, which is eternally present.

So we have to learn how to let go and ease into the motions of life, especially when we become erratic or let our focus drift from where we are now. This is the birth of discipline, where one learns to concentrate and let go - allowing the stream to take us to wherever it will.

What is the nature of discipline?

Before we get into the nature of discipline, which is a really wonderful experience, we must let go of our previous definitions of it. Discipline, to many of us, is what we give to children or adults who are being punished because of poor behavior. Or it's something that's handed out to condition those who don't abide by "the rules." Yet both of these interpretations have nothing to do with the actual meaning or origin of the word discipline. By its literal translation, a discipline is "something that we practice for it's own sake." Yoga is a discipline. Tai chi is a discipline. Prayer, Meditation and visualization are all disciplines. They're the things we do for their own nature, in order for us to learn to be here now! When discipline is forced upon us, we're not experiencing life's true nature. We're only experiencing someone or something elses inability to accept change, which plays off of our own inability. The Zen Master striking the student or shouting at them, though effective in bringing the student back to the present moment, is actually the result of the master sacrificing his own focus to interfere with someone else's wave. Though the master's role or intention is to help the student find concentration in the now, they lose their own inward focus in the process. So this idea of discipline being a practice or method of training to wake people up to the rules, laws of nature or society, has transformed from "the act of completely focusing on one experience," into a method of conditioning everyone into one standard experience or norm. Thus it's inherent purpose is lost for many of us.

To further this dilemma, the majority of us who do take part in practices of discipline, look at it in completely the wrong way. For instance, we have to recognize that whatever discipline we're performing, whether it's the psycho-physical activity of yoga, the art of tai chi, meditation, etc - disciplines are performances which tend to be radically different than our normal human activities. They're awkward in comparison to our typical human behavior, and usually anything that's outside the realm of normality in adult(and child) life is frowned upon or laughed at. To counteract this natural resistance, we should approach discipline in the same way a child approaches playing in a sandbox for the first time. They're curious at first, but then slowly that curiosity turns to delight. Such delight is the birth canal for complete acceptance. Thereupon the experience is expressed by the full acceptance of the individual.

It's important for us to become aware of our natural apprehensions for experiences that exist outside our norms. They're merely the result of our resistance towards change. After all, we're all fighting change in some way, because we're placing our experiences in a perceptual polarity of having and not having what we desire. This causes us to do things for only half of their nature, which is usually for the nature of good. In fact, how many of us meditate, pray, or do yoga because we believe that somehow it's good for us? I can't think of how many times I've heard people say to me, "yoga is so good for you!" It's an overwhelming majority of us that make this mistake and herein lies the problem... Disciplines have nothing to do with anything that's good for us! They, as stated, are the activities that we do for their own sake.

When we do something just because we believe it is good for us, we're not only resisting the complete nature of the exercise, but also attracting what we don't want. Thus if we apply good to any one thing, we must meet it's bad, because we have to experience waves in their entirety. Any time we apply a particular polarity to life, we must experience both aspects of it, because there's no such thing as half a wave!

So every time we practice yoga, meditation or prayer because we think it's good for us, we're putting pressure and expectations on the experience when there's no need to. We all must realize that there's no such thing as good yoga or bad yoga, nor can one "do yoga" or "not do yoga" - these are our human misconceptions and expectations. There's just yoga, and one's participation in yoga. The same as there's just meditation or prayer and one's participation in mediation or prayer.

By doing something as simple as applying expectation, we're trying to control the nature of the discipline. This is why most people experience failures with yoga and meditation, because they're resisting the nature of the exercise. Or, when we pray, many of us fail to receive the deliverance we're wishing for, because we're trying to control the outcome of the energy we desire.

You see, by trying to force our perceptions into any one thing, we're simply acting out the same fear of change that we're wishing to overcome. For example, if you pray for

Love in your life, you're basically reinforcing the idea that you "don't have Love" in your life. Likewise, if you're praying for rain, the rain will never come. Accordingly, if we continue to perceive life in the same way - with the same ideas or expectations, life will continue to give us the very same energy that we're wishing to let go of - ie. a lack of rain or Love. This brings us to Einstein's definition of insanity, where he said, "insanity is doing the same thing over and over again and expecting a different result." Here we find that our continuance in forcing our expectations on our experiences, is actually what is causing our dilemmas - it's not the experience itself. And so when Confucius said, "A man who understands the Tao in the morning, may die with content in the evening," basically he was saying, "when a person recognizes their expectations are causing their dilemmas, they can let go and freely experience life for what it is."

The Tao, which is a philosophy for understanding the nature of now, simply reiterates time as a function, which doesn't solve anything. It doesn't correct our patterns of behavior, or our philosophical issues - We do! And so when we practice a discipline we mustn't perform it for the purpose of any one thing, like it being good for us, or, because it's what we should do! Nor should we expect that it will lead us anywhere, because how can we be led to where we are already?

The words join, junction, union, yoke, and yoga - all these words are basically derived from the same root. So when Jesus said, "for my yoke is easy and my burden is light," he basically was saying, "my yoga is easy because I have no expectations." The opposite to this expression would be the sanskrit word "ayoga," which is the feeling of disjunction or separation, or, what psychologists would refer to as "alienation." Fitly we never hear anyone express their desire to participate in ayoga or anti-yoga. So if someone's yoke or yoga is easy, they're implying that their union or joining with the eternal now is simple - it's concentrated. Therefore concentration and focus are actually the more appropriate associations for the word discipline, because they're states of being centered.

By this understanding, if we practice a discipline for any other purpose than us being in union or concentrated with ourselves now, we're completely missing the point. Discipline is performed for precisely the same purpose as "knowing thyself" - it's just what we do. Subsequently, that's understanding the Tao in the morning, or being aware of yourself in the beginning of any one experience.

As a result, your union is easy, because there's no separation from you and now - they're the same experience. So when any particular experience ends, you end it with content or as Jesus would say, "easy," because there was no expectations.

By this realization, any discipline performed becomes a complete experience of and for it's own nature. Thenceforth, we're no longer fighting the feeling of being separate from ourselves or the exercise, because we're in union with it.

Disciplines are practices we perform for us to focus on being present now - That's all! Any and all realizations that come after the fact are not a result of the discipline, they're a result of us being aware of life's many torrents of waves in the present moment. Correspondingly, it's always in this moment when we're awakened, because it's the only moment that ever exists! And so likewise we understand that life is eternally now! However we must become aware of it first and then learn to settle into knowledge, which is why we participate in our disciplines.

So let go of controlling the outcome of your discipline and just experience it for what it is. Let the current bring you to wherever it must go! Through this measure, "life is a discovery," because the present moment is when everything new and exciting occurs. By living for the discovery of the new, we're able to focus more intensely on the union we have with ourselves now.

Choosing the Right Cup

One of the most obvious to the trained mind, but unnoticed to the untrained, is the keyhole for an inward philosophy - "the heart is the way!" Yet it's not something that's expressed fully enough, and realistically must undergo a thorough investigation. The reason for this is rather simple - We can assume all we want and take other people's word for it, but then we're falling back on beliefs. Our knowledge is always a derivative of trial and error, or "seek ye shall find." Truth on the other hand is only found through one's own faithful analysis of Life's many wave functions. Therefore, through our examination of life's incessant need to keep moving, we're able to uncover the greatest of wave generators. Therefore what we eventually find, or, come to terms with in a logical sense, is that the direct source to and of Creation is through our very own heart(anagram for Earth). For example, the heart is the seat of our emotional intelligence. As a matter of fact, it has it's own intrinsic brain with some 40,000 neurons(the equivalent to any part of the brain in the cranium). In addition, it has no higher authority mentally, emotionally, or physically, which means the heart is superior in all forms of physicality as well as innate mental and emotional functioning and intelligence. Even still, knowing what we know now from the evidence found through the field of neurocardiology, we find the heart's true way to be something that's largely subconscious, especially when we choose not to explore our relationship with it. In the philosophical sense, further exploration is performed through correlation - meaning, in order to understand the heart, we need to relate its functioning to other phenomenons.

What does the heart remind us of?

By one interpretation, a heart's inception is very similar to the creation of a star. In point of fact, the same as the human embryo collapses in on itself within the womb, so does subatomic matter to form a star. When the process of involution reaches it's peak, the heart erupts into evolutionary existence the same way as a star's formation. Furthermore, both heart and star stabilize and provide for whatever form of matter is created in the field around them.

We find a very similar phenomena when we scale back our focus from this solar system, and view the functioning of the galaxy. In the center of our galaxy, as we've discussed, there's a hub of luminosity, which radiates light and life into its spiral arms. This center is created by the same process as the formation of a star, in that, much larger clouds of galactic dust and moisture fall in Love or are attracted to each other, and then bond to form one very dense point. When that point can no longer hold in this Love or energy, it bursts outward and stabilizes itself the same way as a solar system. Therefore, the luminosity at the center of this galaxy, can be referred to as, "The Heart of the Milky Way." And furthermore, this heart provides the environment for many more smaller expressions of itself - ie. solar systems, and the human heart.

The heart can also be interpreted as one of the 7 cups of creation, which has also been referred to as the 7 chakras or lotus flowers. Each cup or chakra, energetically speaking, is a miniature black hole or channel to the quantum multi-dimensional field of existence, also referred to as the "planes of existence." When pursuing a spiritual practice and/or way of being, we find the exploration of these cups or chakra points to be important for our spiritual growth - due to the knowledge and energy received when they're activated. These activations captivate the individual in a manner that's beyond physicality, and resonate more with the emotions experienced when witnessing an epic sunrise. So in the same way, as a lotus flower dips below the water at night, when the light shines on it in the morning, it emerges from the liquid to bask in the Sun's rays - illuminating it's inherent purpose(to invigorate the onlooker). And since the conscious mind is photonic light in motion, when we become fully aware of these energy points - they blossom like the lotus flower. Therefore, when one begins their spiritual or inward work, they explore the significance of each cup in order to decide which they'll drink from most often. We know this quite well in Christianity, when we think of the cup of Christ, which we'll get back to shortly. For now let's explore some examples...

There are many different practices to awakening the dormant energy that exists within our physical and energetic bodies, which has been described as "kundalini." This is in reference to the serpent like energy that spirals up our spines and invigorates every aspect of our being with the feeling described as "grace." As stated, there are many ways of reaching this experience. Some require rituals, some require mediation, others invocation, etc. In fact, when we look back through our history, we find the ancient world littered with vague translations of connection to these various chakra points, or, cups of life - all for the purpose of experiencing kundalini or any other interpretation of divine spiritual connection(like grace). For example, in Taoist philosophy, when one practices Tai Chi - through a series of motions they imagine energy or "chi" coming in from above and below their bodies, thus allowing their solar plexus chakra to blossom. From there, they move this chi around their body to activate the other points. Chi, which is considered Universal energy or prana, in this case is manipulated from the part of the body that's located in the middle to upper abdominal region, which is also where the Gut brain resides.

Sex has been practiced as a ritual ceremony for the activation of the sacral or sex chakra since as far back as can be measured. In fact, in ancient Egypt, high priests while performing a sexual ritual, would circle what was called an "ankh" over the top of their head while orgasming, because they believed the sexual energy which pulsed out of their bodies from their heads, would circulate back into their bodies energy field.

In tantric philosophy and practice, the physical orgasm is considered the far off echo to the bliss we're all seeking. However, as a consequence, if the physical act of making Love is not cultivated and expressed through the higher chakras, it becomes a part of the path of indulgence, which we should all know by now is the path of "temporary relief." Society today is mainly based off sexual rituals and worship, which are designed to trigger indulgence. Just turn on the television or open up a magazine and pay attention - we're mainly being sold on sex, which is a disgusting and primitive exploitation of something that's inherently beautiful. Hence we're also a bandaid culture, or an indulgent society only seeking to relieve ourselves again and again.

In many occult and spiritual traditions, dating back to ancient Egypt and on to India, connection to the Pineal Gland was considered one of the most sacred aspects of meditation. It has been referenced as our third eye(Ajna), or, the eye in which we see beyond the third dimension. The gland itself, is located in the center part of our brain - directly behind the brow bone between the eyes, and is responsible for the release of the serotonin derivative melatonin, as well as the psychedelic chemical dmt, which has been said to awaken deeper levels of consciousness. These chemicals are vital for the regulation of our perceptions as well as our circadian rhythm or biological clock. Therefore this gland, which is on average no bigger than the size of a rice grain, is extremely important to our physical and emotional being.

The french philosopher, mathematician and scientist Rene' Descartes even referred to this gland as the "principal seat of the soul," or, the place in which the soul resides during the period of life. And so, practices or disciplines that initiate connection with this gland or third eye, one could say, "is for the purpose of activating a higher form of being existing within us."

At the same time, it's also where many practice in the art of astral projection and various other concepts of Universal connection. Thus the pineal gland many believe to be the point in which we connect with our higher self, which has the ability to travel through the cosmic dimensions - revealing the great wonder that exists beyond our 5 senses reality.

All of our chakras hold significance to our overall being. However, there's one cup that remains pure in the sense that no other cup produces the same energy or effect. It does not trick us, nor does it lie - like many fools claim. In fact, it's the place in which we see life beautifully, and in harmonic patterns. It's our source and keyhole to inward truth - the cup that Christ, Krishna and Buddha would have drank from - The Heart, which we've established is an anagram for "Earth." Earth translates back to the name Sophia, which means - "wisdom." So, one could say, "this is the place where we find the Spirit of the Earth, which is the essence of wisdom." But why this one? Don't the others bring us wisdom(knowledge of truth)? The answer simply is this - "in life we must choose wisely."

The Gut Brain is part of our lower conscious physical self, and operates our fight or flight mechanism. It's also responsible for digestion and various other bodily processes which keep human beings in a state of desire and indulgence. So by connecting to the gut brain or solar plexus chakra, we're activating the deeper parts of ourselves which bring us anxiety and hunger. Combining these two feelings or emotions, we slip into the more pronounced state of insecurity, which is none other than a lack of fulfillment. Now, the blossoming of this chakra may be associated with "power," however, it's only a temporary fix. Power, as described, is a product stemming from the desire for control, which is an illusion, because the physical life is a temporary experience. In the end, everything must be let go of, because Creation must change itself in order to continue discovering itself. It's not to say that this chakra doesn't play a significant part in the overall energy of the human being, but more, it should be understood as a symptom or after effect of connecting with something larger.

In many ways the Solar Plexus chakra is a chakra in polarity. After all, why would we have disciplines which target this area? We do this because of the a desire for fulfillment, which is the result of unfulfillment. So the activation of this chakra indulges our need to feel full, which is a powerful feeling. However, any polarity where we're living with the expectation of fulfillment, will typically end in us reverting back to it's opposite, which is unfulfillment. That's why this area of the body is where we find hunger in a polarity with satiation. Our feeling of being satisfied corresponds equally with its opposite which is hunger. Thus, connection to this chakra is wise for the sense of fulfillment, but unwise if it's the source of our fulfillment.

Sex or the physical act of making Love, though considered a holy or wholly experience of creation, has become an artform that's greatly misunderstood. Unfortunately, this misunderstanding prevents us from experiencing it's true expression. You see, human beings mainly relate sex to a desire, which revolves almost entirely around lust or indulgence. This of course is a direct result of material mindedness - where we perceive life in terms of fulfilling physical needs. By this measure, lust is no different than a craving or deep yearning for something that we don't have. So if we're lusting after life, we're most likely experiencing a state of unfulfillment. Therefore our understanding of sex has changed considerably over time. Whereas it was once a powerful form of spiritual connection, it's now a predilection for feeling temporarily satisfied. And as mentioned before, "sex, or the physical orgasm achieved from sex, is only the far off echo of bliss." An echo of course is not the real thing. Just like the wake of a ship is not the ship passing - It's only an idea or memory of it once being there. The same goes for sex, because the orgasm is a fleeting experience. For this reason, by the time we settle into the sheer thrill of it, where one is delighting in the feeling of being alive... it's gone - the ship has passed! Hence, participating in sex to experience fulfillment runs the risk of us living off the echoes of a greater experience, which makes it an illusion. Be that as it may, our recognition of this idea runs the risk of us following an even more dangerous path...

Since sex is the sacred union of male and female, many believe that we should abstain from it and only use it for the purpose of procreation. This however is merely playing into a polarity. After all, what is the opposite of indulgence? That would be abstinence. Therefore by choosing to abstain from having sex, many end up torturing themselves into eventual indulgence. Here of course is where the guilt comes in, or the error that many mistake as "sin." To be clear, sex is sacred, it's not a sin! However, like all sacred experiences, it must be cultivated or treasured for its inherent purpose - which is union. So on the other hand, having sex just to get off runs the risk of indulgence spilling into other areas in our lives. And by indulgence I mean giving into our insecurities, which stem from our fears. But on the other hand, abstaining from indulgence simply delays the profundity of our need for union, which will eventually force us to indulge.

Sex is wonderful if it's understood, and careless when it's not. Nevertheless, in order to understand or find yourself centered or in union with it, you must search far inward to see it's deeper meaning, which we'll get into towards the end of this book. For now, I want to reiterate that abstaining from sex for the purpose of not wanting to indulge in sin - by way of a religious sense, is completely inhumane. Yet it's also inhumane to look at it as a form of physical relief, because of how divine it actually is. And so just like the solar plexus chakra, the sacral chakra is wise to connect to for the sense of fulfillment, but unwise as the source of fulfillment.

The Pineal Gland is one of the most difficult for us to break down, because it's considered by many to be our source of insight or intuition. In this case, we're referring to our mind's ability to imagine the deepest possibilities for any particular scenario, and in turn, be able to acknowledge the root of it. By connecting to this Ajna chakra or 3rd eye, we're given a glimpse into the astral reality, which is equivalent to the entire field of perceptual awareness - existing beyond our window into the spectrum of light. It's incredible to realize life's many facets, particularly to our levels of mind. To have a mind or even the neurological contraption we call a brain, that's able to center itself in the midst of the endless expanses of the Universe, and in turn start measuring the whole thing, is quite extraordinary! The pineal gland is a point of recognition, if not the inner-view to that expanse, and respectively - a key source of our measuring. The best part is not having to be conscious that we're using such a device - we just use it. And one has to imagine or even wonder, "what if I could master how to use it?" Therefore, we find the pineal gland at the forefront of many of our disciplines. But before we condone our pursuit of its mastery, there are certain aspects to the word "Pi-neal" that must be broken down for its further understanding.

PI(π) is in reference to the ratio that's been associated by mathematicians as the ratio of God - $\pi = 3.141519 \rightarrow \infty$(infinity). This could be a direct reference to our 3rd eye or inner-eye, which connects us to the infinite. After all, this is at that center of that neural contraption we just mentioned. And if the human brain is simply mimicking consciousness at the highest levels of the cosmos - then we could say that this is our central brain - or "at the heart of our conscious brain." Likewise, if you knew that the infinite could be measured and or found within a pathway inside you, wouldn't you want to explore it?

In the historical context, we find the same curiosity. The third eye is referenced as a source point to the infinite or pathway for PI, and is found all over the world. In fact, several religions and many of the mystery schools praised this numerical expression because of its ability to translate in nearly every aspect of life - whether it be in nature, biology, megalithic architecture, or even language(it was infinite). So we have Pi - the ratio for the Mathematical God, and then we have "neal," which can translate to "kneel." So to put the two translations together, the pineal gland would be the place in which we "kneel before the Infinite or God." However, something remains a miss to this equation, because we're 2 chakras away and off the mark from our true center - the heart. And if the brain in the cranium naturally plays second fiddle to the emotional brain within the heart, we know we're definitely not starting in the right place. This is why connecting to the pineal gland as the source of our spirituality is all but another trick of the spiritual world. It would seem obvious for us to go straight to the pineal in order to connect to God or the Universe, and that's why it's a trap. Since when have we ever been disconnected from the infinite?

When we look deeper into the math surrounding PI, we find it in direct correlation to Cronus or Saturn in Greek and Roman mythology, Odin in Germanic mythology, El or Osiris in Egyptian mythology, Tammuz in Babylonian myths, Yaldabaoth in the Gnostic mythology, Allah in Islamic theology - who is in relationship to Yahweh or Jehovah in the other Abrahamic traditions, Shiva in the Hindu traditions, and also The Lord or Satan in Christian theology. These gods or deities are known for their governance of judgement and order, trickery, chaos and destruction, death and resurrection, as well as time/space. When we look further into the nature of these beings, we find many interpretations referencing a similar cosmological if not celestial story.

To give an example, most all solar systems found in our galaxy, are binary systems(two stars), which makes our single star system a galactic anomaly! However, it was said that Saturn(Cronus or Kronos), who was the brother or partner-star to our Sun(Hyperion the Titan in Greek mythology who fathered the olympian sun god Apollo or Sol invictus in roman myths), never fully came into itself - resulting in its destruction or failure(but not before it terrorized the solar system with an unpredictable set of gamma ray bursts). From this supernova, a deluge of light(massive ultraviolet pulse or wave) enveloped the solar system and blacked out the sky(due to the massive electromagnetic frequency which caused major volcanic eruptions, superstorms, and an ice age when it resonated with the center or iron core of our planet). Cronus or Saturn, as it is in Greek and Roman mythology, is said to have swallowed up the planets(his children), excluding his son Zeus(Jupiter), who is Ishtar or Isis in other traditions. After this celestial event, and when the skies cleared, the planets not only seemed to be shuffled or shifted into different orbits, but the light of Saturn appeared far off and was dimmed greatly. Here it was said Isis(Jupiter) the wife of Osiris(Saturn), wrapped him in a swath(halo) to be dressed for his journey to the world of the dead(underworld) - where Osiris would reign. From here we also see a correlation in the Olympian, Zeus(Jupiter), who imprisoned his Titan father Cronus(Saturn) in Tartarus with a bronze gate or veil(the rings of saturn), where he would remain indefinitely.

In light of this brief depiction of our history(which should be further investigated by the reader), it's easy to say that we have wonderful mythologies across the world, but they're essentially stories for children to understand our cosmology. Furthermore, it appears they're all telling a similar tale... A celestial body, which once was the key electromagnetic force for our solar system(perhaps center), collapsed or was destroyed - resulting in a deluge from the waters(light or electromagnetic frequency) of heaven. The solar system was enveloped by this great event, and reformed itself with the Sun(Apollo in the Greek myths and Zoe in Gnostic ecology) at it's center. Saturn, because of it's loss of power or charge, remained as the "black hole Sun" or failed star in the outer part of our system - ever present and watchful. This watchfulness is in reference to a particular anomaly.

On Saturn's south pole we're able to observe quite an unusual phenomenon. The south pole of Saturn has a circular vortex(rather than the hexagonal vortex on the north pole, which is in reference to the greatly worshiped 6-pointed star) resembling the Egyptian "Eye of Ra" or "All Seeing Eye," which is found in much of the Egyptian and occult literature around the world. Consequently it's also something that's found through many of our current philosophical and spiritual traditions - for a direct reason. For instance, in Saturnian traditions, which are still found to be the undertone to most religions today, The Lord(El, Satan, Cronus or Saturn) is ever present, ever watching, ever judging, and ever waiting his return to the celestial throne of our solar system. This is the storyline we hear in the abrahamic religions, as well as the eastern religions(particularly hinduism and buddhism). Even more, when we break down religions, as mentioned, we find them in relationship to certain astronomical events. Thus it's no surprise that roughly a thousand years from the apocalypse(great change), Saturn(The Devil) will flip its poles and be released from its rings(pit, gates or chains). This is referring to a specific astronomical event expected to take place approximately 1,000 years from now - Will this failed star experience a "resurrection" through a natural process?

In conclusion, we find a remarkable story about extraordinary celestial events in our skies, which seemed to have taken place longer than 6,000 years ago as some Christian sects claim, but more like 12-15 thousand(the cause of the ice age) or even millions of years ago(extinction of the dinosaurs?)... Be that as it may, and as we all know, time is a function that can be debated very easily, which unfortunately keeps many of us divided. We can all agree that something happened in the past, however, because the details are so extraordinary, differences in opinion become the grounds for greater conflict - resulting in our further division. Therefore, Saturn, which is a character that's either embraced or greatly rejected in our theology and mythology, remains a controversy in most intellectual circles.

However, through this character or planet - Saturn, which is heavily associated with PI or our 3rd eye, we find a consciousness that's not only controversial, but is in many cases the opposite of the Love we're wishing for(even if certain sects claim otherwise).

Now again, it's not to say that this Ajna chakra should be avoided(nor should Pi or Saturn), because it's special in it's own right(the same as all religions and schools of thought). However, it should not be the beginning, middle, or end of the ultimate, but merely an expression of the larger puzzle. In the same way the mysteries schools were nonbias in their examinations of light and dark arts(which play off each other) - so should be our examination of the Pineal Gland, which is our insight(in-sight) into the realm of infinity. Yet on the same hand, what we find through its evaluation - the eyes are grown as one sense and an aspect of something larger! Thus our in-sight, again, is only a tool for spiritual expression, not the true source.

The heart, at our center, maintains the point which pulsed us into being and will continue doing so until the day we reach the great waterfall. Then, naturally I would imagine - like all stars or oscillations, the energy from this central point will collapse back into itself, and erupt or radiate back into the quantum field(the most finite field of matter). This idea relates perfectly well with how we're created... When in the womb, well after the embryo has been fertilized, the very first sign of life is when a heartbeat emerges from nowhere(now here). From that pulse, the heart stabilizes its oscillations or pulsations and begins to build the rest of our body(this process of creation ironically takes about 7 days and happens during the 5th week of pregnancy). The heart grows the brain stem, which grows the cranial brain(home of the pineal gland). In fact, the heart grows the emotional part of the cranial brain(amygdala, thalamus, etc) before the rational brain(left and right cerebral hemispheres with 4 lobes), which actually grows out of the emotional(we are emotional beings before we're rational). The heart then grows the gut brain, which is the home of the solar plexus chakra(we are emotional and rational before we're physically developed). The sexual organs, which house the sacral chakra, then grow out of the physically minded gut brain(we are emotional, rationale, and physical before we're sexual). So we find the heart as the source which perpetuates the growth of the human fetus, and further onward - the growth of the human being.

The heart also produces the largest electromagnetic field in the body - radiating 5,000 times more powerful than the field produced by the cranial brain. In fact, depending on the average human beings state of emotion, the heart can resonate a field with a circumference of up to 10-12 feet - which means we're living in a direct energetic relationship with our immediate environment. It was said that people could feel the presence of the Buddha up to 3 miles away. Which of his chakras do you think was resonating that frequency? Jesus was described as being bright like the Sun. Why do you think he was referenced in this way? Buddha's presence and Jesus' brightness both relate to electromagnetic frequencies that are characteristics of a star, which correlates with the pulsations and frequency of the human heart!

The heart is what regulates our electromagnetic flow as well as our electrochemical flow(it pumps 2 gallons of blood per minute and over one hundred gallons per hour through a vascular system that's roughly 60,000 miles in length). Even more, it's the source of our emotional intelligence, which we know means intelligent energy in motion.

Furthermore, we can understand the heart as our source of creation, because it's the singularity or point in which mind, body, and spirit are the same experience. In relationship, it's also the place where we find understanding in the Tao or nature, as well as where we experience union or yoga.

Think of the chakras as points of awakening, or points to the hierarchy of being. Some might even say our being cannot exist if one chakra is missing, because they're all necessary aspects of the human experience. Yet when we become the seer and the doer, and are participating in the experience, there's always a beginning or starting point, because life has to expand and contract from a central point to come into existence - this is oscillation. The lower chakras(Solar Plexus, Sacral, and Root) are consistent with gravitational effects and indulgence. They're the energies of the body which keep us grounded with the earth. The upper chakras(throat, third eye, crown), are considered the non-physical chakras of the mind and spirit, which are levitational. The chakras of the mind and spirit keep us connected upward with heaven, and the chakras of the body keep us connected downward with the Earth (as above, so below). Therefore human beings are a perfect balance between gravity and levity. And just like all dual forces, there's a point in the middle where we have balance or union - in this case with heaven and earth. This place is also where the seeds of consciousness are planted. These seeds or ideas of being, root themselves as beliefs within our body(subconscious mind), and further - grow as knowledge or wisdom within our cranial brain(conscious mind). We plant these seeds in the good soil found in our hearts, which is the plasma field maintaining our physical existence. Thus we've rooted ourselves in the Earth, in order to grow towards the heavens - just like any other plant or organism on this planet. The heart, as we can now understand, is the balance between the effects of mind and body; gravity and levity; conscious and subconscious, and is thus able to produce what Jesus referred to as, "good fruit." These are just some of the examples of why the heart can be considered the wisest cup.

So when seeking to find yourself in the "Now Principle," if you're focusing on your breathing, then you're focusing on your lungs, which are in the same area as your heart. This should come as no surprise when we observe the fact that our heart and lungs are in a symbiotic relationship - The lungs regulate the heart beats, and the heart beats regulate the breathing pattern. Therefore when you focus on one experience, you're really focusing on the other, because they're representative of the same larger experience - life and death in every breath; life and death in every heartbeat. And it's from this relationship of inhalation-exhalation, where we find resurrection or renewal in corresponding intervals with life and death. So if you focus on your heart when we breath in, and imagine yourself going inward to the source of all life - you are experiencing what sparks Creation. Likewise, when you breath out, and imagine radiating outward from that point - you are experiencing Creation's need to change. Furthermore, when you breath in, life comes back into you, and the process repeats itself. This is a complete expression of being - an inward and outward experience. Hence, the wise man or woman will always drink from the right cup if they're intuned with the larger experience, which is represented by their inward experience. Ie. Pulsations of the human heart, as well as our breathing patterns, are in correspondence with our galactic pulsations.

The Rhythm of Life

When we think of life in terms of necessity, we can narrow those necessities down to four simple things - light, food, water, and air. Well, when we examine these forms of sustenance, we find that our needs for them are unequal. Although every one form of sustenance exists as a different wave that nourishes the experience, their effects and necessities have different magnitudes. For example, human beings can go long periods without sunlight, up to 40-50 days without food(in some cases living without food), a maximum of 3-8 days without water, but we cannot go more than 3-5 minutes without breathing. That means that breathing is the most fundamental part of our physical beings functioning! Pay attention to your breathing for one minute... How many breaths did you take? On average we take between 7 and 10 breaths a minute depending upon our state of being. If we're flustered or over-exerted, we may take up to 20 or 30 breaths in a minute. When we're sleeping, and our breathing is deep, we may take 3-5 breaths a minute. That means that our breathing is not only changing rhythm constantly, but is also extremely important for our well-being. Strangely though, we only take our first breath the moment after we're born(we didn't need to breath in the womb?). In fact, as soon as we exit the birth canal and enter this gaseous environment, a doctor is usually there to slap us on our bottoms - forcing us to breath. However this is - for the majority of us, the only practice we'll ever receive in breathing. Why is that?

    Breathing seems to be the most overlooked aspect of our species, even with numerous schools of thought teaching wonderful philosophies on the subject. It's almost unthinkable, knowing what we know now, how we're all not mastered in our breathing at a young age. Yet again, the things that seemed to be the most logical, are rarely figured out... That being said, when we do educate ourselves on the matter, and put a discipline into practice, we're capable of learning a great deal. Everything we've discussed becomes apparent, and further, we're able to perceive life in a much grander way. Thus, our correlations with this process expands to levels never imagined, as we witness our most basic functioning realized as a profound measure of the entire Universe...

    The process of inhalation-exhalation is a fundamental life expression, and can be described in terms of electric and magnetic principles. The action of the in-breath is the cause of electricity. It's also the generative principle of life which accumulates and grows our being endothermically(internal generation of heat). It's the gravitative principle of attraction and cause for evolution, or, the action which attracts change. Thus with every breath our cells grow, regenerate, and or evolve. The in-breath can also be considered the male principle or the seeding principle in life, because oxygen is taken in and deposited all over the body - planting the seeds of life. It's also the electric and creative principle, as well as the assembling principle that produces form, because the inbreath is causing countless electrical reactions - restructuring our bodies internally. On the opposite end, the out-breath is the cause of magnetism. It's also the radiative action or the exothermic reaction of dispelling heat. The out-breath or exhalation is dissipative or repellent - the outward force. Exhalation is deconstructive and is the action of disassembling form. Therefore with every out-breath we lose a bit of ourselves, or outgas energy that has been transformed into carbon-dioxide(in-breath of plants - giving life to our environment). To relate the two, the in-breath or electric life principle is the winding of the cosmic clock, whereas the out-breath or magnetic life principle is the unwinding of the clock. So you see, breathing is an electromagnetic process which represents the becoming and unbecoming aspects of the Universe. It's inward and outward, attracting and repelling, more and less, charging and discharging, allowing in and letting go, endothermic and exothermic, male and female, etc. The universal pendulum swings toward solidity or form with each inhale, and toward formlessness with each exhale. These alternating swings are referred to as "oscillations." Hence, because you breath - you're oscillating(altering of magnitude or position around a central point). With each and every completion of the breath we find life and death, which means, "in every breath we live and die," or, "form and change."

The breath also regulates the rhythm of the entire body, on every level. Therefore, if your heart is beating fast, typically you'll experience a heaviness of breath. If your mind is racing, usually your breathing will become erratic. When we're calm and relaxed, we find a similar breath. In fact, if we observe our constantly changing day to day interactions, we'll find our breathing pattern shifting with every wave of change. How about when you take a deep breath and let go of a long exhale? Then how do you feel? Hopefully you're beginning to see the bigger picture and how important the breath truly is. Breathing corresponds to the rhythms of our lives!

Even more, when witnessed, breathing is a wonderful philosophy - a Universal philosophy. For every one inhale - there's an exhale. When we breath in - our bodies expand, when we breath out - our bodies contract. The Universe itself is a field of expanding and contracting energy - an ocean of rising and falling waves. Our breathing is synonymous with that same nature. It represents not only our state of being, but the state of the entire Universe, from the Galactic down to the subatomic.

Henceforth, practicing or participating in a breathing discipline provides us with a platform for experiencing many wonderful things. It also has the potential to guide all humanity to the places and or states we're wishing for and likewise, deserving!

We're all in some way desiring the experience of salvation, deliverance or "letting go," are we not? This state ironically is found through our exhale. It's the other half of the artform that inspires us to continue onward(which requires our letting go from moment to moment). As a result, when we breath in, we find the place of form and creativity we're also wishing for. And so, we're all creatively coming into and letting go of life simultaneously, which is a complete state of inhalation-exhalation.

Ultimately, and to drop the needle back on a familiar tone or to play the broken record again - there's only one place, state, or moment that this can happen, and that's here and now - the only place that can be experienced. Here and now is when and where the whole Universe is inhaling and exhaling each breath in corresponding rhythm, with you!

Breathing; a technique, practice, or discipline of breathing, is simply meant to make us aware of our expanding and contracting energies in this present moment. The reason for this is quite simple - separation is an illusion created by the in-breath, and the out-breath awakens us to such separations and illusions. This is all a function of the greater illusion brought forth by the total wave - the entire field of being or One Image. Your recognition of this idea, or awareness of it, is the experience of total oneness, which is considered the ultimate experience in buddhism. It's when we come together with ourselves again and recognize that we're just as much the Universe as the Universe is Itself. Know however, this moment of omnipotence does not make us God, but rather gives us the feeling of complete nothingness - an exhaling of form and rejoining with the greater cosmos. Yet this feeling is only momentary, because when we breath in, all form comes back to us creatively, which we must assume is the very nature of this Universal emanation.

Life is therefore a back and forth and in between game, in which we're all seeking balance. However, it remains the simplest experience one could ever ask for when we take in the most obvious point, which is that it's always happening now! It's one continuous wave - One moment; an inhalation and exhalation of electricity and magnetism; form and formlessness - coming on through You - the measure of the entire Universe...

So we practice the discipline of breathing to focus on what we are doing in the present moment, because in this moment, we're experiencing what the whole Universe is experiencing. This is of course is what it means to be made in the Image of God, because you - the focal point of Creation, are living a reality to prove your own measure. This being expresses its fulfilment and it's complete nature through the rhythm of it's own breath. So breath not just because you must. Breath because you're willing to experience the complete nature of yourself.

When you find your true rhythm, you'll experience Nirvana, or, a letting go of the breath; a letting go of form. But on the other end, you'll also experience enlightenment by breathing awareness or electricity back into yourself. And thus, when you're paying attention and completely aware of yourself - you're moving with the current - the wave of the One moment, which is rising and falling; inhaling and exhaling, in order to know and let go of knowing itself. Even more, you're breathing is what perpetuates all of it...

The Master's Breath

When choosing a breathing discipline, we must find a method that works for us as individuals. However, we also need to understand that breathing is a product of the same wave nature that perpetuates the whole Universe. So the technique that we eventually settle with, should mimic our individual understanding of that nature. What I've found to be the best breathing technique, at least for me personally, is what's referred to in pranic healing as *The Master Breath(7-1-7-1)*. This simple method, when mastered, is capable of revealing a tremendous amount of wisdom. Furthermore, you can reset your whole life to it. It'll become the rhythm that you live by. And it goes like so:

1. Sit or stand with your back straight
2. Close your eyes and breath through your nose
3. 7 seconds inhale
4. 1 second holding the breath
5. 7 seconds exhale
6. 1 second holding the exhale
7. Repeat the process

Some other things to consider... A quiet environment helps, but it's not required for focusing on breathing. You're going to want to imagine that a string is holding you up from the crown of your head to the ceiling, allowing your spine to stretch and relax in a straight position. You are also going to want to keep your tongue placed on the roof of your mouth, completing the energetic connection in the front of your body.

The inhalation process or 7 second in-breath, allows the emotion of life to enter into you. The one second hold with the breath, is to allow yourself to experience one moment of complete fulfillment, enlightenment or creative form. The 7 second exhalation or out-breath is the Nirvana or letting go of the breath, which is allowing change into your life by contracting inward. The one second hold is experiencing a new beginning of formlessness, or birth of a new cycle of being, which is exactly what emptiness is.

The Master Breath in it's entirety, is about experiencing the full nature of the cosmos, both life and death, light and dark, expanding and contracting - the completeness of everything that is happening right now. It's a beautiful philosophy, and a rhythm that allows the wave to flow in one pulse with the larger current. Even more, you'll learn how to focus your energy in the same way. It is a lifelong discipline that promotes a mindfulness of being.

Chapter 6

The Field beyond the Breath

We're reaching a critical point in the timeline of the anthropos, which has been prophesied for some time. Though many of these prophecies are the product of our collective consciousness' self-fulfilling beliefs, and perhaps the interference or manipulation of outside forces, none of it really matters in the end. In this Universe, or even to narrow it down -this galaxy, existence is a product of constantly changing torrents of energy. So any one thing that is, must change in order for the greater galaxy to continue coming into itself. Our part in the story line is small in comparison to the grand scheme, however, our effects will send large ripples - echoes that will reach far and wide. We must understand that we're moving into an amazing direction - regardless of what the Abrahamic religions have crammed down our throats about the judgement of our species by Allah, Yahweh or Jehovah, whom the Gnostics have labeled to be a demented alien posing to be our Creator(that's not to say that this challenge isn't absolutely necessary for the experience!). We're actually on the brink of a great change in consciousness - a time when humanity will have the potential to raise its frequency or luminosity, and fully connect to and support their planet, who has been referred to as a divine torrent of energy or "aeon" named "Sophia."

However, in order for any one thing to reach such heights - challenges must be set. Our challenge has been rather unique in its unfolding, and I would imagine has been watched closely by the spirits of our galactic spiral, perhaps even higher consciousness beings who are awaiting our graduation. Imagine what that day will be like? Thus our challenge must be a rather important experiment for the cosmos. Overcoming the nemesis, one who doesn't exist within the normal set of senses, is a quite the task! But that interval is coming to an end - our wave is rising to a crest.

Part of the nature of the anthropos, is learning to experience cycles or waves in their entirety. For humanity, we've purposely fallen as highly spiritual beings into the realm of the materially minded, which is dominated by separatism and debilitating polarities. Be that as it may, as all intervals, spirals or ellipses reach the height of their challenges, a certain point or opportunity arises when they can expand their consciousness and move into a new realm. This particular part of the human game or experiment, is the "apocalypse" that's been discussed throughout our history - all things must come full circle in order for intervals to reach their completion.

Yet, words like storylines, during the course of this game, have been co-opted time and time again. Under the circumstances, the word apocalypse no longer carries the same meaning as it once did. In fact, when we break down this word, we find two root words existing within it - epoch and ellipse. Epoch means a period of time in history which typically marks a notable event or certain characteristic. Ellipse means a regular oval shape, traced by a point or focus, which moves in a plane, so that the sum of its distances from two other points is constant - it's a symmetrical oval pattern. So the word epoch-ellipse, or, "apocalypse," actually refers to the completion of a revolution around an oval pattern.

In our case, this is in reference to galactic time, or, the precession of the equinox, which describes our solar systems designed course through the galaxy. During this course, the solar system is expected to complete a 25,920 year revolution through the galaxy, which has been studied all throughout our history. In fact, in order to know "precession," the plotted course or changing degree of the stars would have to be studied almost every single day for at least 2,160 years! Furthermore, our planet only moves 1 astronomical degree every 72 years. Therefore, this was a highly studied astronomical science, which would have to have incited a person's imagination as to what this precessional direction implied! From this need to explore the purpose of this experiment, precession has been interpreted in many ways.

By one interpretation, this completion of a galactic interval or year can be represented as the end of the 12th zodiacal age. Yet it also is inherently expressed as the beginning of a new interval! This contridictive point in our timeline has been therefore prophesied in two separate directions.

In relationship to one standard - it's the end of the world. By the other standard - it's the beginning of a new world. Well when we look back in history, we find contemplation of both ideas. However, what many tend to forget is the "emergence factor," which references the dawn or beginning of something new, which requires the end of the old.

In astrology, this period has been referred to as the dawning of the age of aquarius, which represents a cleansing of not just the physical, but the mental, emotional, and spiritual realms(this takes place at the beginning of a new galactic year or revolution, which is signified by a period of 2,160 years called Aquarius). By the mythological view, it's the golden age, or time when life is golden and bright; when humanity comes into harmony. In the religious view we look to two arenas for its interpretation - Christianity and Hinduism.

Both religions are similar in quite a few ways, because they're both seeking to cultivate something deeper within. At the same time, they're both struggling with a psychic or psychological nemesis, which is what ultimately distorts their view of reality. However, as we know, the interpretation means everything. By the hindu view, the dreamstate or "manvantara" as it's called, is coming to an end. In light of this, we open our eyes and enter into the "prolaya," which is the awake period. The total wave of this experience is referred to as "maya," which in one way can translate to "illusion," or in the other "creation" - its total experience. Therefore, the hindu view is claiming that humanity is waking up from their illusions, overcoming their psychological dilemmas, and finding out exactly who they are. The christian view has more of a polar outlook of chaos to order. The chaotic aspect is the "Revelation" - or time when the deeds and actions of mankind come under judgment or "under review." Yet from this period, arises the Christos(Krishna in the hindu view), which christians tend to believe is a singular being. However, Christos, which means "anointed," dates back long before Christianity, and was referenced as something else entirely.

    In fact, the gnostics claim the Earth itself was anointed in its infancy by a massive energy torrent named Christos. This consciousness or aeon as they call it, originated from the pleroma or galactic center, and enveloped the planet - allowing Sophia the opportunity to balance and redeem her inability to cope with physical incarnation. So by this interpretation, Christos is something singular, however it's quite large in its expression - large enough that it could blanket the Earth. And if this has happened before, what makes us think this blanket doesn't still exist? In fact, if Christianity is looked at from more of a science fiction perspective, it's possible to say that the Christos anointed the Earth again during Jesus' time frame. Thus, the story itself might be in reference to a much larger event that took place. Perhaps, if this massive torrent of energy made a connection with Earth 2,000 or so years ago, judging by the distance between the galactic core, it might take that amount of time for this energetic torrent to reach us fully. This of course is in reference to the "second coming of Christ," which appears to be more of a galactic phenomena. The outcome of this event seems to hint that a new frequency will become available for this planet to merge with. This higher consciousness, one can only speculate, seems to be a much larger event, rather than a singular being saving the world. Why else would the christian interpretation reinforce the idea of Christ returning in the sky? Perhaps Christ is singular, but more of a galactic being comprised entirely of a massive field of energy, similar to the beings described by the gnostics who exist in the galactic core?

    Furthermore, it's plausible to think, at this point, that energy is consciousness, but in order to bring energy from one point to another - a connection needs to be made first. There's a name for this in physics, which is called a "birkeland current." This is usually referenced as a set of currents that flow along geomagnetic field lines, which connect the earth's magnetosphere to its higher latitude ionosphere. However, it has also been described as any field-aligned electric current in a space plasma. In our case, astrophysicists have determined that the solar system, particularly the Earth, is connected to the galactic center by such an electrical current. This means that there's a direct electromagnetic channel that connects us to this pleroma or galactic center referenced by the Gnostics. Therefore, if the Christos is an energy torrent that originates, like Sophia, from the galactic core, such a current would be needed in order to anoint or connect one to the other. Also, and more importantly for our sake, the arrival of that energy would be something incredible. Think of a gamma ray pulse or a massive current of ultraviolet light, enveloping the solar system(perhaps why the Vatican's "Lucifer telescope" is pointed directly at the center of our Milky Way?). Might this event cause great devastation? Maybe... However, frequency is frequency, and considering that ultraviolet light or luminosity is an extraordinarily high frequency, the subatomic matter or conscious pool of information in this solar system would be immensely affected, and most certainly rise to exist on the same level of vibration as this torrent of energy.

Either approach, whether it be the religious point of view of Christianity or Hinduism, or even the anti-religious view of Gnosticism, all have their merits, because Christos is represented as a redeeming or awakening quality, which can mean several things. By one definition it can mean compensation for the faults or bad aspects of something - this can relate to "saving." The other definition is to gain or regain possession of something, in exchange for payment - this can relate to a pact or deal. But again, the gnostics refute the redeemer complex, which originated in ancient Palestine, because it refers to a singular expression of one being saving all mankind. But this is our history unfortunately - many different avenues to investigate, and typically, as we find, all great things are co-opted and turned into something they're not for the purpose of control. We can only speculate at this point, but fortunately we have a great deal of information to sift through in or to formulate our own opinions.

I personally see the Christos in more of a cosmic view - a blanket of higher consciousness that will at the very least envelop the Earth, perhaps even the entire solar system - providing humanity a channel for higher spiritual vibration or development.

This idea of the entire world being redeemed can be explained in many ways in regards to the spiritual nature. The scientific approach, however, is still - in my view, expressing the same idea, because of its pragmatic approach towards the energetic qualities of consciousness. For now though, I want to shift gears and focus on the emergence concept, which relates quite well to both the christian and hindu theological perspectives.

A Period of Emergence

The period, referred to as the consciousness of Christ or Christos, is an expected period and long awaited aspect to our cosmic cycle. In this era or age, all humanity will experience a change, beginning at the quantum level and spiraling upwards. It will not only change the way we view ourselves and our relationship to the greater Universe, but will also guide us into a new and perhaps more harmonious direction. This higher dimensionally minded approach, is one that merges the concepts of philosophy and science, to form one universal ideology that reflects the original vision of the anthropos.

Now to reel back the ideas of conspiracy, which actually translates back to the latin "conspirare" or "to breathe together," I want to reiterate that this universal ideology is geared toward the cultivation of "co-creation," not world domination or destruction. We have to realize that we're essentially a species that has been co-opted by a parasitic consciousness that feeds off of our mental energy, and whose goal is to dominate, suppress, and or destroy the human genome completely - the "Nemesis Challenge." This is not the original storyline meant for the divine experiment, but again - an anomaly that must be overcome. When it is, we'll breath new light into our union with the Earth, and be able to coexist with nature in a matter that has not been achieved in the divine experiment thus far. Further, we'll be able to remove the barriers between our separate natures and come into ourselves completely. By this idea, spirituality will be no different than physicality, in that their understanding will be universal in processing.

To find further meaning in this concept of emergence, we must look back in our history and give thought to how someone might have come into this knowledge.

Initiating the Emergence Principle

In the true initiatory realms, nature or Sophia revealed herself to an initiate in the form of wisdom. However, this wisdom was not received from her directly as we might think. In fact, our current understanding of divine conversation seems to always be through a physical means, like an angel coming down from the sky or a talking burning bush in the desert. But these are merely material-minded interpretations of divine inspiration.

Nature, which holds the spirit of Sophia, would reveal wisdom to an individual in a rather unique way. This is why Gnostics and the Mysteries were devout in approaching inward cultivation through natural principles. What they were able to determine, through what seems to be generations of paranormal and psychic research, is that coupling natural principles with one's inward frequency or self, would open a pathway to the quantum-spiritual realm of consciousness. This would allow the individual mind the opportunity to discover things that were once foreign to their understanding. By this interpretation, knowledge is not gained through divine intervention, but through the familiar concept of an "original idea" or "private revelation." After all, considering that the anthropos is a divine experiment, overt interference would be against the core concept of our species natural evolution, which needs to be kept in the enlightening state of wonder or discovery. Therefore only by connecting to one's own heart rhythms could the initiate unlock Sophia's potential. This mechanism was purposely built in as a guidance system, so that the anthropos could have a relationship with their divine mother, as well as the cosmos.

The idea of a built in connection to the divine realm is not so far off from what many schools of thought in both philosophical and theological arenas have proposed. In fact, we find throughout our history many examples of human beings connecting to nature, and in-turn - enlightening themselves spiritually. Two of the most common examples continue to fascinate and inspire humanity today. And although their preferred meaning relates almost entirely to a religious ideology, we can still extract a great deal of truth from these stories.

For example, in Buddhism, during Siddhartha Gautama quest to let go of the pains of the material world, one day he decided to sit under the Mahabodhi tree, which is considered a very old and sacred fig tree in Bodh Gaya, India. It was under this tree where he decided to participate in his famous meditation. During this time, he remained still, allowing bugs to bite him, and birds to nest in his hair. Meanwhile, within in him - the mental game continued, as Mara, the buddhist interpretation of Satan or "Evil one," hoped to lure him into having selfish thoughts and visions of his very beautiful daughters, whom he'd left behind. However, his goodness prevailed, and overtime he was given profound insight into the illusions that befell mankind. So after 49 days of this initiation, while in deep meditation, under the light of a full-moon, the morning star appeared in the eastern sky, and he became an enlightened one or Buddha. After this remarkable experience came to pass, the Buddha stood up and professed his gratitude to the tree which gave him shelter. From that point on, the tree was known as the Bodhi tree or "Tree of Enlightenment."

This story tells a remarkable tale of not only a man overcoming his inner dilemmas, but the defeat of manipulation and mental attack from an outside force. These two themes are found throughout our history in many tales. In fact, this story is quite similar to Jesus's 40 day initiation ritual in the desert, or barren land, in which he defeats Satan at his own game, resisting his temptations and professing his undying devotion to his true Creator. When Jesus emerged from the desert, with the help of John the Baptist, he immersed himself in the Jordan river. In his emerging moment, he was illuminated or anointed with the power of the Christos or Christ.

Both the quest of Siddhartha and Jesus have remarkable similarities, in showing humanity's power to overcome the psychological game of manipulation brought forth by evil, which is the reverse of "live." These stories are also showing something that's unfortunately overlooked or misinterpreted by both christians and buddhists today...

Nature is the undertone in both stories, and plays a significant role in each character's enlightenment! Though it's easier for buddhists to agree, because of the importance of the Bodhi Tree, most christians refuse to look at the variables because of their devotion to the materialistic interpretation of the gospels or good news(which never really got out). By this idea, the story is not questioned beyond its usual religious interpretation and reflection to otherwordly measures.

Yet, when we look close, in both stories, nature is the guiding principle, because each character is left to its mercy. Siddhartha sits under a tree to weather the elements, and Jesus wanders the desert or barren land. Each setting describes the archetype of man vs. nature, which in most cases is grossly misrepresented by the outcome of "man overcoming nature." Yet the clue that leads us to the importance of this realization is quite subtle and very easily missed.

In the case of Buddha, when he becomes enlightened, he thanks a tree? And Jesus, goes for a swim? From these two rather comical and interesting questions, we find that one's enlightenment or first act as an enlightened being, involves an interaction with nature. Why is this?

Well, the key to this is in the interpretation of both stories. The tree gave Siddartha shelter, and the water cleansed Jesus's soul. So in both cases, Nature has assisted in their enlightenment. The tree, which is the sacred tree of enlightenment, relates to the tree of life or immortality found in Judeo-Christian theology, and is symbolic for the Holy Spirit of the Earth. And as we know, once Jesus takes a dip, the holy spirit comes into him. The water of course is symbolic to the waters of life in Genesis 1:2 - "And the earth was without form, and void; and darkness was upon the face of the deep. And the Spirit of God moved upon the face of the waters." This state of formlessness is the same as the state of desirelessness sought by Siddhartha, and that which Jesus achieved by denying the temptations of Satan. So the act of the baptism, is to fall beneath the waters of life - into formlessness, and to emerge clean for the "the light" to envelop the initiate. Thus when Jesus emerges from the water, the Holy Spirit enters and anoints him with the Christos.

It may be hard for many to consider, given our religious bias, as well as the notion that nature is a simple creation or expression of an extraterrestrial God. Yet, this interpretation is truly the product of conjecture by institutions who are not willing to give up their dwindling authorities. How many stories do we find around the world today where a human being discovers truth through nature? And how often do we find those stories corrupted and turned into something they're not? Why is it that our species always has to be left to the manipulation of higher authorities and not be allowed to find truth in their own way? Unfortunately, this is a dilemma in and of itself. We're taught from a young age that we're flawed, and suffer original sin or error. Further, and probably the most disgraceful, is that we need to devote ourselves to outside forces, in order to be free of this burden of being human...

It's really sad to think that we're not strong enough to rip up this inhumane doctrine built upon the insane idea that one should feel guilty for being alive, and see that the emotion of our anthropos is something very real, as well as divine in its origin. However, this is our challenge, and in light of it being just that - a challenge, we have to know that we wouldn't have been faced with it if we weren't able to overcome it! This is what stories throughout our history continue to tell us.

The anthropos or the human species, is remarkable in it's ability to overcome sometimes impossible challenges, as well as great suffering, and in turn make the best of their situations. That's the true spirit of humanity, and even more, the story of Sophia - who's intention was for the anthropos to be the "measure of all things." Yet as we know, an error created something diabolical that would challenge her creation indefinitely.

At this point in our timeline, we're on the precipice of waking up, and seeing nature for it's true beauty. Even more, realigning our allegiance with her, and fulfilling the divine measure of the anthropos.

Our emergence is one of poetry and wisdom, and the cliffhanger to the storyline which has yet to come to pass. Be that as it may, as these great changes envelop the Earth, through a natural cosmic order and the Love of Sophia, or, philosophy of the anthropos, we're quickly finding truths which are undeniable. These truths are not only affecting our way of thought, but are transforming our beliefs about the world in which we live. In time, which is rather trivial in the grand scheme, we'll develop more of a universal understanding of nature, and all it implies.

To give valuable examples of this, we look to two areas of science which hold major breakthroughs in the way we see not just the macroscopic universe, but also the microscopic universe.

The Wisdom of Sophia Comes Forth

In quantum biology, which has been underscored since its inception in the mid-20th century, major breakthroughs have been made in the field of energy transference and interaction. Quantum biologist have established the flow of electrons in our body to be the most intrical part to not only our energy resources, but also our overall state of health. Originally, we looked at electrons as a negatively charged fundamental to electricity, but when science discovered that our awareness or observation actually brought them into physical existence, it realistically began to change how we saw the world.

What they found, was that particles at the subatomic level(particularly electrons), not only behaved in the strangest of ways - working through various levels of time, function, and space, but were also a key ingredient to our biology. This means that electrons function as a primary source for the bodies electrical energy, which we can confidently say is consciousness in motion. And just like a battery, when a positive force is applied to a negative charge, we get an attraction that creates an electric reaction. This shines light on why being positive simply makes us feel better. Is a state of mental positivity actually creating an increase in electrical connectivity within our bodies?

On the other side of the coin, the universal cosmology is about to change from a theoretical mathematics model, to the electric universe model, which is guided by modern day plasma physics or as it has been recently described, "Plasma Cosmology." This move into mainstream plasma physics is a continuation of the older electromagnetics model, which pioneered much of the technology we have today, and is basically considered electromagnetism 2.0. This is another key ingredient to the understanding of our biology, because it reinforces the fact that electricity and magnetism are what charge and create the macro and microcosms - "as above, so below." So it doesn't matter if you're looking at an atom in the body or a star in the sky, because electricity and magnetism are the inherent forces of both - working from the smallest scale to the largest levels of existence. Hence, the universe is electromagnetic again, as it originally was in our past understanding.

This full circle interpretation goes even further, and actually shines light upon many of our spiritual concepts of metaphysical energy transference, which are practiced in traditions around the world today. However, as it was once classified as hocus pocus or magic, it can now be explained by the electrical relationship between the human body and nature. This new understanding gives credence to Arthur C. Clarke's 3rd law, where he states - "any sufficiently advanced technology is indistinguishable from magic." Therefore, what was once overlooked as science fiction or metaphysics, has left the realm of fantasy, and has become a provable or at the very least experimental science - confirming Clarke's 3rd law.

In yogic traditions, "prana" is considered to be a universal energy, which is received from three major sources: the Sun, the Air, and the Earth. Each source produces its own form of prana that varies in size and complexity. Solar prana, which is considered the most finite, is a form of energy that operates at the quantum or subatomic level of existence. Air prana, which is considered a medium grade energy, operates at the molecular level, and is said to be moved similarly with the ebb and flow of the ocean, which is influenced by the Moon. Earth prana, which is the most coarse in it's design, operates at the cellular level, and is taken in through water and electrical energy in the ground and atmosphere. All three major forms of prana are essential for our energetic well-being.

In theory, when one participates in yoga or other forms of focused meditation, they're bringing prana into their body through several ways. Solar prana or spirit prana, comes in through the top of one's head or crown chakra, which is considered the universal pathway for consciousness(also comes in through direct contact with the hair and skin). Air prana, is naturally absorbed through our breathing patterns and filters through our chakric channels. Lastly, Earth prana comes in through the soles of our feet, travels up our legs, and into our root chakra at the base of our spine(can also be absorbed through water).

By focusing one's intention, this prana can be absorbed at greater levels - allowing the body to charge and recharge itself - increasing our overall energy and sense of well-being. However, what we've been missing in these practices, and what will ultimately bring the rational and emotional minds together, is a scientific correlation found between physics and metaphysics. This factor existing between the two fields, is what quantum physicists stumbled upon in the early 20th century, but were unable to further calculate for its more direct implications. Quite simply, these subatomic particles are major properties in the elevation of our life force. Furthermore, we have natural interactions with these elementary wave-particles(a clever expression for the subatomic plasma that is neither particle nor wave) unknowingly, all the time.

The sun, which is a massive organism, or star, is composed entirely of hydrogen atoms that are bonded, compressed, and fused to form a level of helium. It also produces an electromagnetic field that reaches far beyond our furthest planets. And as we all know, the Sun is in a symbiotic relationship with the Earth, which needs its light in order for organic life to flourish. Though what many are unaware of, or do not focus upon, is the Sun's photonic nature.

Photons(quantums of light) are emitted by the sun in massive quantities called gamma rays. These gamma ray pulses can be described as high frequency electromagnetic waves. Each wave, which pulses out into the solar system, has a stable vibratory signature. This stable frequency was called the Schumann resonance(7.83 hz), and for quite some time was the frequency of the solar system. However, after 2012, the Sun's frequency has increased dramatically, and has fluxed between approximately 10 and 30hz ever since. The change in frequency has been associated with the period of space we're entering into, which is of a much higher frequency. This provides astonishing implications in itself. For one, what happens when the subatomic frequency of the solar system changes, particularly to the physical structures existing within it? And two, how does it affect the consciousness of that system?

People have looked to the Sun for their knowledge and energy for quite some time. In fact, Sun worship is found throughout our history, but has been labeled Pagan, occultic, or even heretical by our religious institutions(go figure). Yet, knowing what we know today, we have to wonder if these ancient cultures were expressing a deeper knowledge of the Sun, which was lost in time? The emerging sciences today would have to lean strongly toward yes.

You see, even more important than the changes in the Sun's frequency, is the realization that it's sending electrons, as well as a small amounts of protons and other light ions to our planet in massive waves called "solar wind." These electrons, as stated, are a finite source of our electrical development. For instance, why do you think human beings need sunlight? It's not just for vitamin D! Why is it that sunlight makes us feel so good? To answer this question we're going to take a look at something closer to home, which involves the Earth itself.

The Earth is also an organism, but unlike the Sun, is created of nearly every kind of element that we know of. In fact, all of these elements combine to form a planet that is unique in every sense of the word. No matter where we go, we find life - even in the most inhospitable of places. Needles to say, Sophia is special in her own right, because she has the innate ability to create things that appear to be uncreatable, and make it look effortless for them to exist - She is gifted to to say the least. Be that as it may, there's something deeper about her that we touch on time and time again, but usually come up short as to why. This part of her is one of the fundamentals of life and is considered among several others to be a life giving force which makes this human experience possible. We've discussed it briefly with our breathing patterns, but now we're going to explore it from a more scientific perspective. This requires us to take a look at our blood.

Our blood is about 55 percent plasma, which is about 92 percent water. As we all know, water is a wonderful conductor of electricity if it is either mineralized, charged with electrons or ions, or infused with a degree of salt. Well, considering that almost 20 percent of our blood is sodium, we can confidently say that the water in our bodies is charged, which would allow electrochemical information to be passed through our bloodstream quite efficiently. And since we're between 40 and 90 percent water depending on our age, we know water to be the source which conducts electricity in the body.

Well, like us, the Earth too has blood! Some have claimed this blood to be oil, which is completely inaccurate. Water in fact is the blood of the earth! And so we see that the Earth moves much of its water through ground currents within its crust - similar to blood vessels in the body. Just like in our blood vessels, electricity, which can be defined as the flowing motion of charged particles, is passed through these ground currents, making the Earth not only an electromagnetic field generator like our Sun, but also an electrochemical organism.

These ground currents move electrons to places all over the Earth. In fact, human beings have been connecting to these ground currents for quite some time. This is why we "ground" the electrical boxes in our houses and buildings. By doing so, we're tapping into an unlimited supply of electrons, which is equivalent to the negative terminal of a battery. Yet there's another more interesting way that human beings can connect to this free flowing current of electrons, which has to do with taking our shoes and socks off.

Being barefoot actually connects the body with the Earth's electrical current! Yes, in fact, being barefoot has proven to not only stimulate "good vibes," but also charges our bodies with electrons. Thanks to quantum biology, we can observe the effects of increased electrons in our bodies through grounding, as well as observe its various points of electron reception. Some of these points are known, others are not. For example, many of us have experienced "static electricity." This event is typically the product of our hair follicles catching electrons. And since we're also radiating a positive charge, when the positive charge or proton meets an imbalance of negative electrons - we get a static charge. This usually happens when we touch other objects, because every object holds a certain charge - positive or negative. So when the hair follicles of our skin, which are lubricated by sodium mixed water molecules coming from our pores(sweat), come into contact with another object - if there's a buildup of electrons, our hair follicles attract to the protons existing on the other object. The result is a neutralization of the imbalance of positive and negative particles, which gives us a shock or charge. Yet the body not only attracts electrons on its surface, it also takes electrons into itself.

One way we've established this is through our in-breath, which is an electric principle of human expression. The outbreath of course is the radiating or positive principle. Beyond the breath and skin, we also attract electrons and pull them into our body through energy points. This of course is in reference to our chakras.

Since we're electrical beings, our bodies regularly produce positive charges due to our radiating principle, which can be associated with body heat. The Earth's surface is electrically conductive, because it maintains a negative charge due its free supply of electrons, that are continually replenished by the atmospheric electrical circuit(the atmosphere is teaming with charged particles), as well as interaction with the solar wind. Thus when cosmic rays or solar wind interact with the Earth - they react with the Earth's Van Allen belts, which are part of the outermost region of the magnetosphere. These belts neutralize the radioactive properties of high energy charged particles and in turn allow certain particles to enter into the biosphere where they're further absorbed by the water and ground - due to the magnetic pull of the Earth's molten iron core. Furthermore, the Earth keeps electrons flowing in grounds currents, which are moved by the heat and magnetism radiating from the core itself. And since the soil is almost completely mineralized, electrons basically can move freely through it.

Both atmosphere and ground make the earth an electron supplier which, when barefoot, stabilizes the electric charge in the human body and literally brings balance to our being. This happens through absorption at energy points, chakras or charged points in the body, which are capable of receiving electrons. One point just so happens to be at the center of the soles of our feet!

So by walking barefoot on the earth, we're able to ground ourselves to a free supply of electrons which are extremely beneficial to our well being. In fact, it's said that walking barefoot for 30-60 minutes a day can reduce a person's pain and stress, as well as regulate many other bodily functions. Here are a few things that grounding, or "earthing" as it's been called, do for the human body: biological clock or circadian rhythm stabilization(better quality sleep), post-menstrual cycle(pms) regulation, reduction of the effects caused by respiratory conditions like asthma, glucose level stabilization for diabetes, heart rate stabilization and increase or decrease in electrochemical flow, increase in overall energy, an increase in immune system activity and response, and the list goes on...

The correlation between walking barefoot and positive health should come as no surprise, but it does, and we have to wonder, why?

When we investigate this question, we find an interesting thing... Wearing shoes is a rather new trend! It's only emerged in the last 2,000 or so years. But even still, much of the footwear was of a leather sole up until the 20th century, which still allowed an electrical connection with the Earth. Today, almost all footwear is with a rubber sole, which is void of electrical connectivity. Sleeping in beds is also a rather new experience in terms of our species history. We slept on the ground through much of our past, which meant that we were in direct connection with the Earth almost constantly. Needless to say, overtime, as we've become more "civilized," our direct connection with the Earth has become almost non-existent. To most, it's only a place where we dwell, and therefore this place is only seen as an object that exists beneath us.

However, thanks to this emergence in consciousness, which has allowed our science to become more openly holistic in its processing, we're seeing complex theories turned into applicable practices. For example, we're now able to observe electron interaction with the human body at many levels, propagating that our physical and energetic health begins with these basic elementary particles. On top of it, we're also able to recognize that our awareness influences these elementary particles. Which means, when we become aware of our emotions, or energy in motion, we actually have the ability to move electricity throughout our body - just by our intention or level of focused mind.

If you don't believe this last statement, just try focussing on any part of your body right now... can you feel that particular area? This feeling is the result of your awareness bringing an increased electrical current to that specific area. Are you experiencing any pain or discomfort in your body? Focus on this area and imagine a much stronger electrical charge initiating a healing reaction. This is how people are able to heal their bodies through focussed intention(it takes a considerable amount of practice for most, but in time we can all heal ourselves through intentional awareness).

So to relate the electrical properties of the Earth and the Sun to the properties of the human body, we can see electricity as something which exists at different levels of our understanding. However, each level is vital to our development as human beings. Correspondingly, when we look at the battle between microscopic organisms in our bloodstream, fighting each other endlessly - causing countless electrical reactions, we can understand that the discord at one level creates harmony for the higher. The waves of this micro-organic fight ripple electrically through us, perpetuating larger and larger electrical currents, or, ripples. The human being itself also ripples a current, which bonds with the current of all human beings. This collective electrical current or consciousness, directly affects the electrical current of the Earth, which also is in an electrical relationship with the Sun. Further onward, the Solar system is in and electrical relationships with other systems around ours, as well as with the galaxy, which is in an electrical relationship with the Universe. All of these relationships work from the subatomic level and ripple upwards to the Universal. Likewise the Universal ripples downward through many levels until it reaches the subatomic - everything is in relationship. And without a doubt, we would have to know that prana works in roughly the same way. The three types of prana, are consistent with three separate levels of electrical activity - quantum, molecular, and cellular, which begin with the relationships between protons, neutrons and electrons. You see, "the fight" or "battle" as we've described is merely expressing the polarity between fight vs. harmony. Yet, what we must understand is that the entire polarity is precisely what we would refer to as the "duality of relationships." Each conflict, challenge, or relationship explains the electrical connectivity at one level of our being, which ripples electromagnetically towards the higher. And because we're conscious, we can direct our awareness to bring light or cosmic attention to any area we wish, which brings us back to electrons or prana as they've been referred.

Through our connection to these sources, whether it be directly through grounding, breathing, solar exposure, or through intention based exercise, we're able to increase our electrical connectivity by charging our bodies with prana as it is called in metaphysics, or electrons as they're called in physics. Therefore both fields are referring to the same energetic interactions, which also means that both fields can begin to work together to understand each other. This of course can only benefit our already separate minded species, and bring us back into a more holistic philosophy.

There's a growing number of people who are aware of the constant energetic reactions taking place on this planet. Whether we choose to refer to these reactions as pranic or electric really matters not. The point is, when we become aware of these sources, which are at our disposal, we can begin to not only reconnect to the earth, but establish a consistent flow of electrical energy to our bodies - increasing our overall state of well-being. In this emerging consciousness, which can be associated with a change in galactic and therefore solar frequency, information such as this will begin to become more mainstream. When it does, these simple experiences which are natural occurrences in our lives, will become important to us again.

So, basically what we find in our analysis, is that the same symbiotic relationship that exists between the Earth and the Sun, is happening on every single level, from the universal down to the subatomic - it's all relative. It's all one process, which perpetuates the continuance of motion and function for every one thing that is. The best part - we, the anthropos, can observe its interactions with not just our bodies, but as a process existing throughout the cosmos.

Nature provides us with a tremendous amount more than we give it credit for, which is only because of our disconnection with it. As we hone in our focus and stop searching outside ourselves, as well as our planet for the answers, we'll begin to recognize the true wisdom Sophia is trying to show us. When this happens, our overall cosmology and how we view the anthropos will change dramatically. For now, it's left to the individual to explore nature's endless wisdom, and in turn continue bringing an electrical charge back into themselves, in order to become a lantern in the growing dark - a beacon of luminosity.

Chapter 7

To Believe - And the Machine that Makes it Possible

It was once described to me, "Life is like music" and, "If we understand ourselves the same as we would an instrument - we can play for the birds." It was poetry hearing such beautiful words, even though the man who spoke them is long gone. Yet it really matters not, because the expression was heard loud and clear - cluing me in to the mystery behind language as well as the spoken word. In fact, words are merely hieroglyphs to something deeper - a primal, yet beautiful aspect of the anthropos. If we pay attention long enough, and become more and more aware of our languages, we begin to find certain dormant expressions within them. That's why words are often considered arks of wisdom, because they're preserving deeper expressions of life. Through these expressions, we realize how certain words represent an emotional value, which speaks volumes. Words like Love, hope, faith, truth, poetry, art, Life, wisdom, etc. - all these words are cover sounds for something far more meaningful. When we hear these words, or any particular grouping of words like the phrase above, we're resonating with the roots as well as the growing flower of the Universe. Even more, they make us feel as though we're the electrical conduits for the synthesis, or, music of life. So perhaps my encounter was brief, but the effects were long and loud - propelling me into the beginning of a deeper contemplation or meditation as it's often called. Quite simply, no one ever forgets the first time they hear the "emotional language," nor do they forget how it makes them feel.

You see, Life has a sound that resonates differently with us all - at as many levels as can be imagined. Sometimes I would tell my nephews, "go into a quiet place, listen, and find that sound." And when they responded, "What does it sound like?" I would say, "You'll know it when you find it." My approach may have been cryptic, especially for curious children, but the truth is simple - "we're all, in some way, haunted by a soothing tone that lurks in between Life's incessant whispers." It plays a game of hide and seek with us - "catch me if you can!" Well, not really. You see, the sound we're all searching for is rather simple to understand. Our bodies are instruments! But we would never know this fact until we found ourselves in a state of complete awareness. Only in our fully conscious moments, do we find ourselves plucking the heartstrings - experiencing the full on beauty of life. When this event finally does unfold, it marks the beginning of conscious union with the beautiful subconscious - harmony between all levels of being.

Our beliefs are a lot like strings on a Universal Harp, whereas our subconscious mind is what represents those strings. However, there are many strings, keys, and combinations of tone to this harp, which takes some effort to keep in-tune. For example, if a string is loose, it may need to be tightened. Or if a string has snapped or gone awry, it needs to be replaced. Some strings are even a finer quality than others! Whereas some produce different sounds. Well, if your instrument is in need of tuning, then the music you're playing will not sound right. The music in this case is in reference to your beliefs about life. This is why it's important for us to listen to ourselves - to listen to what that inner-voice is telling us about the frequencies we're putting off or taking in. It's giving us clues to help us determine what strings/beliefs need to be tuned or changed. This can be a difficult challenge in itself. And so it has been expressed that "listening to yourself is a discipline that takes practice." Yet, in time, anyone can learn to understand their subconscious as they would musical composition.

Unfortunately the subconscious mind is greatly misunderstood, and looked at as more of an adversary in society today. It's even considered a paradox or an unknown to many. And because the unconscious or subconscious represents approximately 90 percent of our functioning, many choose to completely ignore it and view it as an entirely separate or alien aspect to themselves. After all, our thoughts give us enough trouble, don't they? Yet what we continue to miss, is that the majority of our thoughts and reactions derive from our belief system, which is housed in the subconscious. So in order to find the root of our conscious dilemmas, we must explore our subconscious mind. However the idea or feeling of separation is what keeps us from its exploration. Therefore, to achieve unity within ourselves, we must first unite our mind, which is both conscious and subconscious.

The subconscious mind represents the roots of our being, whereas the conscious mind represents the growing flower. Both root and flower, as we all know, are not separate from one another. So why is it that we perceive the two aspects of our mind as separate? This question is where we disseminate knower from known, and visible from invisible, because we find the things we cannot see to be typically what get ignored. It doesn't mean they don't exist, it just means they aren't interacting with our sense of awareness - which we usually associate with our physical 5 senses.

The subconscious, or roots of our being, are completely enveloped by an intangible aspect which is closely associated with feeling. Now, touch can be considered feeling or the sensory perception of feeling something, however it still does not measure up to emotion. Feeling is merely the trigger of energetic qualities interacting with a belief or idea we have about something. For example, skin is soft only because we believe it is soft, not because it is. Why isn't soft skin called hard? It's just an agreed upon interpretation of something, which we've specified as fact. Now I'm not trying to convince you to view the composition of your skin differently, but rather show you that feelings can be interpreted differently. In fact, we can interpret feelings quite easily if we try. Emotions however, again, are the deeper part of that intangible something. The reason why we're unaware of emotion, typically is because of how much interaction is going on at one time. Every single one of our billion trillion cells is vibrating an emotional signature. Each signature pairs with others to form larger body parts, which also emit a frequency. Some in fact, like the heart as mentioned, produce a very prominent signature. We can be fully alert and sensually aware of our environment, and never know that our heart was in direct communication with everything within a certain radius around us. However, if we do become aware of it, we can begin to feel more than what our five senses allow us. Yet, someone cannot simply know this by studying it, they have to witness it for themselves, because the subconscious must be felt to be understood. However, this is where we all face another common societal dilemma, which an overwhelming majority of us avoid completely. This aspect, as stated, is the force behind our feelings, which is emotion.

    To clarify the general misunderstanding of the word "emotion," which for the most part is viewed as a mythical beast we continuously do battle with, the word need only be broken down. (E)motion translates to "Energy in motion." That's all! It's simply referring to the energy our beings have in motion. So these feelings that we get(the good one's and the bad ones) are all products of our energetic flow of information. Even more, the majority of this energy is programmable. Meaning, strings can be changed or replaced! The key, as stated, is paying attention; being aware of your 'energy' in motion. What and where is it leading you towards? What kind of state does it put you in? Having an awareness of your body and asking the questions, "What does it feel like? and "what does it remind me of?" will reveal a tremendous amount of wisdom. All you have to do is allow your imagination or conscious mind to show you what it needs to. This is where we see that both conscious and subconscious mind are not separate, but one complete unit. They're 2 aspects of the same one thing that we call Mind. When this realization is made, we experience full mental unity, which is also where we find our principle of mentalism - "all is mind." Bringing this unity into focus takes work, which leads us back to the nature of discipline, where we center our attention upon ourselves in the here and now.

    The first key to mental unity is to understand that an awareness of emotion, coupled with the creativeness of our imagination, will unlock our intuitive senses, or, "Inward Truth." The second key is developing a discipline or method of visualization to work with this inner-truth, which requires muscle memory.
    You see, beliefs exist in the emotional field within the body, as well as the field outside the body. Be that as it may, beliefs are programs which are often triggered by other stimuli. Most often we refer to these beliefs as memories or muscle memories, because we respond in terms of reflex. Basically, when an experience triggers a belief, it fires almost automatically to respond to the experience. Although we're speaking of two fields of belief, both your energy body and your physical body are in-sync as one unit. Yet, we have to understand that the intangible field(energetic) represents 99.99999

percent of everything that is. Which means that if our state of being changes to resonate with a more harmonic frequency, the harp will sound sweeter. Therefore it's your focus and choice of what kind of energy you wish to bring into yourself, which will decide the outcome of how your harp sounds. And here, we begin to awaken a deeper sense of awareness, where we find the courage to face down our inner-demons and tune our instrument to play the sounds we're all dying to hear - call them joy, happiness, or Love.

Do not be afraid of your subconscious, nor treat it like it's a burden. This aspect of your being is beautiful, because it's devoted to serving you, no matter what, for every second of your life. I couldn't imagine a more wonderful gift than this - to be able to have a mind that's dual in its devotion to serving our one true image of being. In fact, we've known this for quite some time. It's a trend found throughout our history. For example, during their 40 years in the desert, on the 6th day of every week, the Hebrews took a timeout from their day to day life to fast and clean the Ark of the Covenant. What is the Ark of the Covenant? Well, an Ark is none other than "a vessel." And a covenant - "a pact or agreement." So simply translated, we have - "The Vessel of the Agreement." The agreement here would be the connection between body and spirit(the polarity of being), with the Mind(also in a polarity with itself) forging the pact. So on every 6th day, the Hebrews would not eat or drink(fasting purifies the blood, which is the conduit for electrical energy), while they cleaned the vessel of the agreement(body and mind). But even more, they would also tune their emotional instrument. In this case, prayer was the discipline and tool for tuning their machine(subconscious). Well, what is prayer? As a verb it means, "To give a solemn request or thanks." As an adverb it means - "A preface to a polite request or instruction." This means that prayer is, primarily, meant to connect with or invoke some larger aspect of something, and secondarily, to instruct whatever that something is into action. So, if your covenant or agreement is between body and spirit, then that means your mind is what's using prayer to link the two together. Yet here's the issue - Most just completely assume that prayer goes directly to God, and they're right to some degree, but not entirely. You see, if your Spirit is a smaller, more focussed and concise version of the One Image or divine pulse that is the Creator or God, then your Spirit is your link to all of it. Thus, by connecting to your spirit or essence of energy, you're awakening your full potential to the One Image. And since we're merely the focused creative image of God, living in uniqueness, then we're also the One Image in it's entirety(Correspondence - "As it is in heaven, so shall it be on earth"). So if we're fully connected to, or, are aware of ourselves, then we're able to command(Cooperative action or mandate) changes and reconfigure our energetic flow. It's the same as consciously correcting your posture periodically throughout the day. Your mind sends the command to the body, and the body performs the action. The Universe itself is the physical, as well as energetic body of Creation. And since you're that very same energy focused into one point, your commands, if coming from outside that point, will be followed in the same way the mind moves the foot. This is how we begin to master our belief work.

First we need to recognize ourselves in our entirety, then we need to center the creative energy of the One Image within us. Once again, it takes practice, which we'll get to. For now, let who you are settle in, and let your true capabilities emerge through thought and feeling. Dedicating one day a week to prayer or belief work, for the purpose of cleaning your Vessel and honoring your agreement, will keep your Harp playing wonderful music.

Mastering Your Instrument

To start, what is it that you believe? And, do you feel what you think you believe in your body? If you don't feel what you're thinking, what are you feeling? We slowly begin to find our negative and limiting beliefs when we observe and ask ourselves these questions. By doing so, our imagination will begin to create images that lead us to understanding. For instance, if you're trying something new and suddenly you experience the feeling of insecurity or emotional pain, ask yourself, "Why am I feeling this way?" Then follow your mind to where it takes you. Usually the mind will create an illusion or image to represent what's causing the feeling. This image or illusion will have a belief behind it. For example, if the image or illusion is a scene where you're curiously trying to grab ahold of fire - burning yourself in the process, then the illusion may have been telling you, "curiosity will get you burned" - this would be the belief. Or perhaps it simply gives a flash image of something, say, "a tree falling down." Examine the image for what it is first, and then how it relates to you. "Do you feel like your world is falling apart? Or do you live with the expectation of it falling apart?" Then ask yourself, "Why do I feel this way?" It's even possible that the mind will show you the root of the belief. For example, the mind may take you to an exact memory, where you hurt yourself trying something new. It's these types of beliefs which are trauma based and hold us back, or, block us from allowing change into our lives - this of course is in reference to what are called emotional blocks or clamps.

The conscious mind is very creative, and it lives to serve you. However, if you're not listening or paying attention to what it's telling you, particularly about beliefs within the subconscious, then it will be difficult to understand. So take your time, and when you're ready, you can work to remove negative or limiting beliefs with visualization.

If you train your mind as you would your muscles, not only can you train yourself to respond with new reflexes, but you'll unlock your subconscious, which is beautiful in every sense of the word. There's no greater devotion than that which is given to you by your very own mind. The very fact that your mind delegates it's own authority, so that you don't have to to be aware of all bodily or belief related functioning, is a relief! The mind simply needs to be tuned like an instrument would be tuned in order to find peace and harmony in life. By appreciating it's inherent purpose, and taking responsibility for it, you can learn to use your mind in a way you never would have imagined.

Yet, there's something else that's important to realize, especially when it comes to life and music... The composer does not live to finish the symphony or to make the end of it the point. The entire symphony is the point - it's a complete expression! This teaches us not to live for the end result of any one thing, but the cultivation of the whole thing. After all, the musician doesn't work an instrument - they play it! When we play with things we're creative, when we work with things we're pragmatically seeking an end result. It is in play when we discover, and through work when we figure things out. And though much of life requires work, we have to see that work and play are in a polarity with each other. So for every one thing we work with, we must also play with something. Your life is a symphony after all - a wonderful piece of music. Cultivate your instrument and play with it - the same as a child discovers the piano. Let the sound of the instrument tickle your imagination - allowing your thoughts to show you what they will.

Within you exists a grand composer - let that aspect come forth so you can discover the music! Only then will see that mind and body, or conscious and subconscious, are playing together all the time. Even more, they're playing through creativity, which filters through our imaginations the same as a wave flows with the larger ocean.

Visualizing the Total Wave

Visualization is more than a practice or technique - it's a discipline of the imagination. Well, what is imagination? It's in reference to "image in action," which is none other than the creative force of Mind. So to visualize, is to be able to create some form of image or reality within your Mind. However, like all disciplines, it requires practice, because the body's mind - the brain, needs time to create it's muscle memory. Does that mean you cannot visualize right now? Of course not! It just means that with practice you'll get better at it, because you'll be able to hold your images longer, as well as take them further. You see, we visualize or imagine because we're seeking not only to catch glimpses of what we wish to become, but also to envision what any particular thing might be. Even more, we imagine because we're expressing our creative principle or impulse. At some point we've all sought to envision the Image of God, have we not? Well, considering that a large portion of humanity has decided that God would be something that exists as a physical being emerging from the sky - we miss that the Image of the Creator is the Image of the total Wave - the entire ocean that is the Universe. So to envision God, we would have to envision ourselves leaving our bodies and looking at the Universe from the outside in. Yet here's the funny thing... If we were to imagine that, then we would be saying that something exists outside this Universe, which would have to be God as well, right? So where does the Image end? Or does it end?

You see, if this experience is one continuous wave of existence, then we shouldn't try to imagine there being an end or an outside. We can only bring ourselves to the furthest most point imaginatively, inward!

I want you to think of a cell in the human body. There's a nucleus, there's a cell membrane, a cell wall, and so on. But on the very outside of this cell, physics shows there to be a field of electromagnetic energy. This field of consciousness is what brings information into the cell, and would be similar to the cell's Spirit. Well, when this cell needs something, it sends an electromagnetic signal to the brain, or, Mind, which responds unconditionally for the cell's entire life span. But this proposes another question... "What if this cell were able to manifest the image of itself outside of the body, and into the bodies energy field to command action within itself?" After all, isn't that where the body ultimately expresses its energetic information? Figuratively speaking and through everything we've discussed, the cell is not separate from this field, so why can't it? Well, because it isn't separate from this field, it doesn't need to... So what makes a human being's relationship with the Universe any different?

We have to know that it's not required for us to go outside of ourselves to receive what we need, because, thanks to Sophia, IT already exists within us. Even when we use our imagination to visualize going outside ourselves, we're not actually going outside of ourselves... We're going further into ourselves and breaking through the blocks that keep us honed in as the focal point! So, if we train our imaginations to take the Spirit of our cells(selves) WAY outside of our physical being - all the way to the energy field that exists outside the physical Universe, we're still existing inside of ourselves. However, we've tricked our subconscious mind into lowering its shields in order for us to command instant changes, particularly to our belief systems. The shield that goes down is the barrier between conscious and subconscious mind. In this visualization, the subconscious mind believes it's one with the Creative Energy that perpetuates all the Universe - the root source. Therefore the commands that are received in this space, are originating from the essence of the total wave - the greater Mind of the cosmos. If you do not Love yourself, you can then command the old belief removed or exhaled/dissipated, and the new belief of "I Love myself" inhaled or generated as the replacement belief. If you do not have patience with others, you can command the old belief removed, and the new belief "I have patience" to replace it, and so on. It's even possible(as I've heard) to perform instant healings from this space, but I would recommend doing your own research on the subject, because energy healing is an art which takes time to learn. Yet, for the purpose of belief work, anyone can learn this process on their own. If you command from this space, it will be done - energetically!

But again, you must train yourself in the discipline of being present. Only in this "presence" can the changes take place instantaneously. And I'll give you a hint as to what this "presence" actually is... It has a billion trillion cells or miniature organisms, living in harmony, creating one unit which functions for the purpose of knowing itself, and it's operating with the same process which perpetuates the greater Universe... You can smile now, because you are all of that, and much, much, more! You are that presence!

From the scientific aspect, which many of us relate to more than metaphysical thinking, we can attribute this process to entering into a theta brainwave state. To understand this more, we have to know that our brain operates at 5 levels of vibration - gamma, delta, theta, alpha, and beta. In this particular moment, or in any normal conversational or higher brain wave state, the brain is operating in beta(14 to 30 hz). In most meditative states, or times of calm and relaxation, we enter into alpha(9 to 13 hz). When we enter into deep meditative states or into the dream space, we're operating in theta(4 to 8 hz). In deep dreamless sleep, when the body is recharging itself, typically at stage 3-4 in our sleep cycle, we're operating in delta(1 to 3 hz). Gamma(25 to 100 hz) is a high frequency pattern of neural oscillation, where it's implicated that the human brain enters into a state where it has created a perception of unity with the collective consciousness. All states of brainwave functioning are necessary for completeness in the human species. However, in a way, not all are created equal for the purpose of creativity. For example, in beta we can consciously think our way through a set of problem-solutions(associated with work). In Alpha we can dispel the concerns or over-workings of the thoughtful mind(relaxation). In delta we can let go of the mind completely, so that the body can restore itself(mindlessness). In gamma we can enter into a unity consciousness(collective consciousness), which is extremely effective in spiritual practice. Yet, in theta, we're in a state where we have one foot in the conscious and subconscious realms(dreamstate). By this measure, we're scaling the wall and can bring conscious thought and action into the subconscious mind. Henceforth, we're able to alter the subconscious world from this space. It may sound theoretical in many ways, but the proof is in the pudding. From this space, you're bringing down the walls of the subconscious, and can convince it that it's in the space of All Creation. And so, you can command Creation to make the desired changes while in a theta brainwave state.

Now to be clear, this will be an energetic change, meaning the belief will no longer exist within the bodies energetic, or, auric field as it's called. So the emotion should not arise when the belief is triggered physically or situationally. However, because the physical belief still exists as a cell in the brain or body, the mind will continue to activate a physical cellular response until the cell is fully broken down(up to 21 days). But, when you realize the emotion no longer exists for that belief, because you don't feel what you're thinking, then you know that you no longer have to react in the same way - the belief is fading. You see, it's the emotion which reinforces the physical belief, because the emotional language represents 99+ percent of the communication in not just our being, but the entire Universe. We're communicating with our environment, or, with life, all the time - we're that Universal Harp after all! Well, if an old string has been replaced, you're no longer playing the undesired tune. At this point it's up to you to acknowledge the sound of the new string! This is important, because if you allow the physical cell or belief to continue to exist, then eventually the string will revert back to what it was before. So, be mindful of the changes you make, and feel the tune of the new string every time the old one seeks to resonate. Eventually, you'll be able to tune your instrument to resonate with higher and higher frequencies of being - becoming a masterful composer in the relationship between yourself and Life.

Chapter 8

Change is Salvation!

To Understand your life as an object, one that continues to take form or shape, we have to consider that before form can do either, it must be conceived by a thought or command. This is consciousness - the very power that evokes thought and action. Therefore, a human being is a product of evolving thought and action. But to complicate the equation, we too create with thought and action! Meaning, there's an interrelated parallel between the creative action that puts us into being, and the creative action that we execute in our day to day lives. This is where we begin to truly understand the nature of the anthropos.

You see, if you're the living image of the thought form - then you are the thought form living! However, like all patterns of thought - you're evolving. This is the Nature of Thought, or call it the Nature of Life. We're so determined to discover the thought-form or the Godhead which creates this reality, we miss the fact that we ARE the thought - in form! Even more, our 5 senses which keep us isolated in a rather small spectrum, make everything real! Therefore, because we have brains with occipital lobes - our eyes can determine what is light and dark. Likewise, because we have temporal lobes for our ears to hear - we can hear. In the same sense, because we have a soft skin which can sense changes in heat, which is a response of our somatosensory area in the parietal lobe - we can feel by the sense of touch. And to top it all off, we have a parietal lobe which also formulates our spacial sensory - meaning we can focus on physical objects individually. This also gives us proprioception, which allows us to be aware of our bodies moving through space. You see, it's you, or in the same way - "I," which brings all of these relationships together! Furthermore, it means life is merely a game we're play with ourselves - "I am measuring all of this, but in order for me to continue discovering It - which really is me... I have to pretend that I'm not It."

When we stop playing into our game of thoughts, we discover that we are the "I," we always thought we were. And basically, we have no idea who that person is. because we've been pretending to be someone else all our lives. Well, that's the beginning of what we call going Inward or "knowing thyself." This of course is a quiet and not so quiet space, because we have a lot to say. So we begin to listen... and we continue to listen... And as time goes on, we peel away layers upon layers of illusions we've adopted through the course of this little game of not being who we are. Until, we come to the conclusions that life just keeps getting deeper and deeper; weirder and weirder; larger and larger. Finally, the moment arises when we realize that we can't micro-manage life - the stream just has to flow. This is when we ultimately let go... We take the proverbial exhale and allow change to simply be what it is - constant("I can't control life!"). In that moment, we begin to pay attention to what's going on now - in the present moment. And we recognize that the "I" we perceive to be, and "the Now" that is, have always been one thing - the same experience. Therefore, living consciously is only putting the "Now Principle" into conscious motion or reality.

By playing into this game, and allowing ourselves to get too far ahead in the future or too far behind in the past, we're assisted in recognizing an undeniable truth... Past and future are only concepts of time - they do not exist! We use them as alibis for not being present, and also as a means of trying to figure out how to control the illusionary experience of not being ourselves. Past history and memories are merely reflections or recollections to the ghosts of yesterday. However, these ghost are simply reminders of what we are missing right now! The future on the other hand is the sight or target that the bowman locks onto. So when their arrow is released, it goes into an intended direction. Therefore, both past and future are merely games we play in order to either remember who we are, or, to decide who we wish to become. Unfortunately, by playing into the past and future game, we actually forget who we are, because we're only alive now! That's why we go through journeys of remembering ourselves, or, our "finding ourselves" experiences, which produce the same result every time - momentary relief for something we already inherently know! Trying to find yourself is the same as trying to find your hand in the dark. You know that you're hand is right there in front of your face, but you physically cannot see it. Therefore, it falls back upon our senses. Since we cannot physically see the Self and or Spirit that creates us, we assume that it's not there and therefore needs to be found. However, that's the same as saying, "because I can't see my hand in the dark, it doesn't exist" - it's all relative. Trying to find oneself is rather laughable knowing that we're always a complete being in the present moment.

Through any process of trying to find ourselves, whether it be past, present or future, we're only wishing to experience who we are through an idea, which exists only as a mental status, or, delusionary aspect of the game. To fully and completely know yourself, which is something that cannot exist any further than who we are in this very moment, we have to let go of the idea that we're not who we wish to be - this eliminates division. And, because truth is always changing, we cannot control the experience of being who we are, because we just are! Why else do we arduously put ourselves through the same routines over and over again? It's not for us to remain present! It's because we're trying to build an idea of ourselves through controlled programming. Routines are simply reminders we set in place for us to remember who we think we are and what we think we're supposed to do!

Yet what we continue to miss, is that these controlled programs keep us fighting the changing current. They simply become, as stated, the alibis which make us hold on to form by way of an established set of thoughts. And thus we continue to miss the greater realization...

We can't hold onto ourselves! We don't have to try not to hold onto ourselves or even the idea of ourselves, because it simply can't be done! One way or another this form has to change. That's the way of Creation - It keeps changing itself. The thought-form just gives the Now a direction(choice), which must then experience the crest and trough of a complete wave.

So if you give life a direction, you must understand that it's only a temporary rising and falling choice for "the Now" to experience. As a result, you're experiencing the wave of life - and while doing so, you can direct the nature of the experience. On the same hand, we have to be mindful of the fact that the wave must complete it's full intervals. At the trough of the wave, or point of change, the wave simply rises anew. Therefore the idea of death is thoughtless when we realize that there's no stop button on the clock. The wave just begins a new interval! It's no different than waves on the ocean, in that there's no separation between wave and ocean. One wave is merely the Ocean's participation in one particular point in It's time. Thus your wave will never part ways, because it's a synonymous function of the greater cosmos - which is always whole.

Right now, in this particular interval, you're deciding what will be "the measure" of the entire Universe. And that's why we say to "be grateful." After all, wouldn't you want the Universe to be that as well? This is also why many believe that "Love is the Way." Wouldn't you want this to be a Universe that is Loving?

During our time in this wave, we're testing our own measure. Which means that if we're the evolving thought-form of the Creator, then Creation is creating scenarios for itself to decide and/or discover what it wants to become. So It created interconnected galaxies, solar systems with stars and planets, organisms and the organic and inorganic elements that make up those organisms, all the way down to quantum particles which then revert back to the greater whole - all of this is created in order for Creation to know Itself! This Universe is unique, in the sense that all of its creations act and continue to create with their own free-will - all for the purpose of reflecting upon the duel question, "How does Creation measure me, how do I measure Creation?" - And it's up to you to decide! It's all a game; a very clever and creative way to pretend not to know yourself, so you can discover yourself! Yet, this idea is nothing new. It's the same evolving thought humanity has been carrying the tune for since as far back as we can remember… "Who am I?" The question is though - "Is it worth spoiling our fun?" Well… it's worth it for me - the evolving thought that I am. How about you? How do you measure yourself? Answering this question will require your thought-form to ripple. Either way, by the time those rippling intervals reach their destination - the place where we all wish to make it back to, well… at that point, I'd imagine it'll all be the same one thing. If that's not what it is already…

So breathe out, and let go of this back and forth drama that dominates us - It's all an illusion of the game; the divine experiment. Bring yourself back home to the present and awaken from the dreamstate. You have always been here! So stop fighting the current of your own wave, and let it take you. Have faith or trust always in Life, Sophia, as well as the greater cosmos. They exist as the higher aspects of ourselves. And furthermore they continue to advance the wisdom required for self-reflection.

When you do let go, and witness the beauty and grace of this wave, that's when you'll release the breath - experiencing the weightless feeling of Nirvana. Then you're saved - you're free! Thus true salvation is not being enveloped by an outside energy and delivered from human bondage - that's not the case at all! Salvation is the experience of letting go the desire to control the evolving measure of your own life, because the new is awaiting your next inhale, which is the spirit of creativity. When you recognize this, you're saved from the trouble that arises when we attempt to fight an unstoppable current. Therefore Salvation, Nirvana, Grace, etc. are all different interpretations of the same principle. And they're all telling us to do the same thing... "Let go!" You have the opportunity to experience that letting go; that Inward Way, right now in this moment and further onward in this moment, until the stream empties you back into the ocean. At that point, you'll have never doubted who you were, because you trusted the cosmos, which has entrusted you with it's very essence - discovering itself through Its own nature, which is a state of constant change.

You are the Way

I often wonder, "why isn't philosophy at the forefront of our minds?" After all, it's a reflection for our "Love of Wisdom!" Wisdom = Knowledge of Truth. So that phrase changes to "Love for Knowledge of Truth." Just weigh that quote by the quality of it's words... Who wouldn't want those words in their life? So when we figure out that the whole world is lost in delusions, we wonder why when it seems so simple. And right in that very moment, you're awakened to something remarkable... "It's only YOU who measures this experience!" It's not your job to measure or interfere with the lives of others, because truth is an individual journey. You can only present a way, and let the rest of the world decide for itself. We can't interfere with each others waves, which is why we're constantly trying to do so. That's why we have to smash particles together at the speed of light to supposedly capture God in particle form(or for whatever reason). Human beings are constantly pushing their boundaries in order to figure out how they can control their reality(because they've been robbed of the truth by political and religious authorities). Hence we've developed a civilization which has little respect for the natural order of things, nor for ourselves. Furthermore, we must realize that If we can't respect ourselves, nor the natural order, then we cannot respect Creation, which means the anthropos will become a failure. Therefore, in the end, our mission in life becomes - "living our own way," which is why the Bodhisattva vowed, "I seek my own enlightenment for the sake of all others." To enlighten of course means to "shine light upon." In other words, he's not awakening or illuminating the world by preaching sermons or giving lectures to people - not in the slightest. He seeks to shine light upon or bring light into himself for others to know, "they too are more than they perceive!" His way is unique in every sense of the word, not just because he's become a lantern in the dark, but more - he's an example of the inspiration or "in-spiraling" we're all seeking union with. The Bodhisattva lives to illuminate the Inward Way, which is a path built upon Universal principles. And his main goal is to live as an example and to show that this path is unique for all individuals. Through this Inward Way, one enlightens their inward qualities, and in turn enlivens the beauty of everything that's outward.

The very nature of your own wave, when enlightened, essentially becomes the inspiration for others. So when we reach a full circle understanding of ourselves, we'll ultimately be living by the idea that one should "Know thyself, but let go of thyself," because the process is becoming and unbecoming - an evolution of the thought-form. And if we're an evolving thought in a conscious pool, we have to know by the very nature of our own mind, how powerful one thought can be! Take the word Love for example. It inspires us to want to feel special for eternity... How far did that thought take you? You see, you are that evolving thought! It was you bringing it there! It certainly wasn't me. And so it is, it is so. It was always you! You have always been you. You were never anything else but you. So let go of holding onto the idea of yourself, because you've always been in a state of wholeness or "holiness."

Furthermore, when you reach that point, if you're not very careful, for a moment you might actually think you were God. Similar to how Freud felt that babies were omnipotent. After all, why wouldn't they? The light of the entire cosmos is coming on through everything that's new! In a strange way, it's plausible to think that children are born knowing that they're here to play this fantastic game of "measuring the Universe." It's a hilarity to think the process which begins around the time we're born - still remains an unfinished symphony. But one also has to wonder if it will ever be complete? Thereupon the question remains... How will You - the One Image in focus, continue measuring yourself?

Life certainly is poetic in the end isn't it? I mean here we are pondering the very point to all of this! Our evolving storyline is a masterpiece of poetry - Universal in its essence...

Chapter 9

The Love that Makes the World Go Round

We're going to shift gears now, and finish off our discussion with something that many consider deeper, and at the very heart of our human dilemma. It's a thought, feeling, and a concept that invigorates all, but something that very few choose to thoroughly investigate. That's why many consider it to be a cop-out answer or phrase that we hide behind. After all, how often is "Love" used as the end all say all expression of human desire and purpose? We throw this word around - left and right, and claim to be driven by it. But do we actually understand it? Can we understand it in our current paradigm?

You see, Love is the true inward dilemma. What is Love after all but a quest to find oneself immersed in the arms of divinity and grace? It's the feeling that spirals around completeness and total assertion that one is alive and full of the very breath that gives life. Love is the endless echoes of our past and the far off dreams of our future. But, in this present moment, right now, what is Love to you?

In order to fully appreciate the gravity of this question, we first must break down it's importance. Love is the approach, search, and definition of completeness in oneself. You may ask the question "Who am I?" all your life, but what are you really asking if we only find ourselves when we're "in Love?" By the sexual means, Love or sex is the focal point for the culmination of two energies entangled together - becoming one. This gives an interesting approach to Love, because physical interaction has the ability to bring us to that ultimate state of being. Yet, as we all know, this moment is, well... momentary! That's why tantric philosophy teaches that sex, or more importantly the physical orgasm achieved from sex, is the far off echo of grace. So if the physical act of Love is all but momentary in it's goal, then there must be something deeper that's a miss. To further perpetuate the matter, we must look further inward to know what constitutes being human.

If by the basic definition of form, we find that one is either male or female, then we know that the human form is in polarity. Man and woman represent the two polar aspects of a human being - the electric and magnetic principles. So by our need to feel whole, we search the world to find the other half of ourselves in someone else.

Respectively, when a woman is interested in a man, she's typically said to be energetically "attracted or pulled to him." Likewise, when a man is interested in a woman, he's said to be energetically "pushed towards her." After all, it's custom for the man to approach the woman, correct? Electricity, being the male principle, is an attractive or pulling force, and female, which is the magnetic force, is a pushing or enveloping force. Thus male and female relationships are synonymous with a push-pull experience, which means the relationship is electromagnetic - they're in-sync with one another.

Yet it's curious to think that, again, we're seeking to find what we're looking for outside of ourselves, when physics proves the whole universe operates in one total uniqueness.

If we continue to call ourselves the measure of this uniqueness, then within us should be exactly what we're missing. That's why we say, "in every man there's a woman inside, and in every woman there's a man inside." Or when a man is requested to be more emotional they say, "you need to get in touch with your feminine inside." This paradox of being a combination of both male and female, really isn't much of a mystery when we recognize and shed light upon the hermetic principle of gender. After All, gender realistically is the explanation of one thing. But just like a piece of paper has two sides to itself, it still is one piece of paper. The description merely becomes a game of which side we're shining light upon.

So in the same sense, by you being the thought form, your consciousness is choosing which end of the polarity will take form. However, just because you've come into form as a male, does not mean you're separate from female, and vice versa. There's still that other aspect existing within you! Consequently, we find ourselves searching for completeness in the opposite sex in order to fulfill what we're missing consciously within ourselves.

By nature of this dilemma, or idea that we need to search outside ourselves to find what we're missing within, we end up creating more conflicts than we can imagine. The reason for this derives from a forgetfulness that the other person is also playing the same game with themselves, and at the same time - wishing to find completeness within us. Yet, if we're all living with different perceptions and expectations of how to see the world, in order to reduce our internal conflict, our desire shifts to change the other person so they become more like us. Herein lays the problem... By seeking to change the perceptions of our missing link or other half, in order for them to become more like us, we're denying the others participation in the game, and thus denying their pursuit of Love. Likewise, we're also changing what it was about them, which made us feel complete to begin with! This is the basic problem in most relationships, which leads us towards eventual defection. And so, here's where we realize the origin of the dilemma... Since conflict begins within us, it really is our own problem! Furthermore, this conflict revolves around our preconceived ideology about not just reality, but Love itself.

Love is the energy that drives human life, pure and simple! As a result, from the time we're born, our parents and family members pour Love into us for the purpose of helping us grow into the same kind of Loving creature they feel that they are. However, what we miss as students in life, is that the Love that's given from parent to child or sibling to sibling is only half Love! It's not complete Love or unconditional Love as it's been referred. We can witness this through a common phrase expressed from parent to child - "you're free to do what you want, as long as you do what I say." So on the one hand our parents give us freedom, but on the other hand they're tying a rope. This is not to say that children don't need guidance, but rather the Love that's given to them should be recognized as incomplete. The reason why this is so, is because our parents too only feel half Love! After all, aren't they suffering the same inward dilemmas as we are? If a parent is not one in their sense of being, or, "in Love," then they're living by a concept of unity, rather than the complete experience of it. So in the same way, we can understand the Love we give our children, as the same that we give our partners - We only give them the part of us that knows how to Love, and in turn seek to instill that same mechanism in the other. We do this so the other person becomes a source to fuel us, because we don't know how to fuel ourselves. Yet again, what we miss, by trying to change the others concepts, we're directly challenging our own. Thus changes in the other will never be enough, because we're only half complete to begin with.

In accordance, we see children acting out the same philosophy that was taught by their parents, and their parents parents, and so on. By our current philosophical understanding, Love is usually only half Love when it comes to our relationships. This is why it takes a considerable amount of work to achieve fullness with others. We can observe the relationship between parent and child or sibling to sibling as relationships that rarely become pure. And by pure I mean "genuine" relationships built upon acceptance and respect, rather than what commonly drives our connections, which are usually based upon a need for the other.

We have to realize that we're all unique individuals, which live by the same relativistic ideas. And because we feel incomplete, those ideas make us judge other people's approach towards us. After all, how else can we determine what is and what isn't if we're not judging the approach of others?

So we eventually see a reversal in the relationship between parent and child, because the child eventually grows to adulthood, and overpowers their parents with their own version of half Love. Now, basically the child calls the shots, and lives by the expression, "You can have my Love, but you have to respect me first." Therefore, they've courageously declared their independence, but in turn have sought to dominate their parents Love with an altered expression of the original - continuing the cycle.

By this understanding, we can clearly see that the inward struggles of our society remain continuous, because we neglect to face WHY we feel incomplete. Why is it that we only feel half Loved?

By the notion that Love is an incomplete concept, we can observe it evolving into more of an addiction. Hence, when a child is neglected by their parents, or feels that they're not receiving the Love they're used to, they go through withdrawal. In this case, like any substance or energy we're experiencing withdrawal from, we become compulsive and agitated. Our behaviors in this state now function with the purpose of finding a remedy that can ease our addiction. As a consequence, children misbehave or they dance, laugh, and or perform silly acts in front of their parents to receive attention. Both misbehavior and exhibitionism, by this measure, are forms of attention seeking behavior - geared toward the need for Love. Thus, childish behavior is often referenced as a need for attention. As a deduction, our parents example of, "you can have our Love(or freedom), as long as you do what I say," eventually morphs into the belief, "I will get Love when I'm given attention." By this interpretation, can we actually be angry at the person in their 20s, 30s, or even 50s who craves attention? After all, that's how they've learned to receive the energy they need! With this idea, we're able to witness the same struggle existing within each of us. Since we're not taught to go inward and complete ourselves, we build generators that only run on a certain kind of fuel. This fuel is the Love or attention we receive from others.

So as we watch men and women continue to play this game, we're able to observe a foundation which is built upon the perpetual desire to fuel one another. The woman seeks attention from her man, and in return the man seeks support from his woman. Support and attention are realistically derivatives extracted from the energy we call Love. So on the other hand, if the man gives little attention, then the woman becomes needy because her addiction's not being relieved. If the man doesn't receive enough support, he becomes aggravated because he too is not getting his desired dose. But we also witness a funny thing happening with this scenario... If the woman seeks too much attention, the man feels smothered or nagged, and thus feels that his independence is being attacked. Yet, when the woman receives too much attention from the man, she also feels smothered and that her independence is being attacked. The same goes for the woman giving too much or too little support to the man. This brings us to the crazy idea that our addictive concept of Love, must be about "balance." And of course, this is completely ridiculous! Have you ever tried to balance an addiction? Eventually all things that operate in a state of balance, that are based upon addiction or need, experience a state of imbalance which is chaos or disorder. This means - no matter what, we have to indulge in our addiction in order to keep it alive. Unfortunately, this perpetual drama is what most of us crave when it comes to our relationships, because we have to constantly express our need for the other.

Men and women are overly dramatic when it comes to Love, because they view it as a means of ownership and fulfillment. This is a material minded expression, because the other half of the polarity is missing... spirit, which is the poetry, or, that little extra that gives spice to life.

Love, for one, is only physical when it's expressed by motion, but even so, it's still a concept we live by because it's of what we're taught in the beginning of life. We have to understand that before we are physical, we're energetic, and before we are created, we're whole. Therefore we only separate ourselves when we put our thoughts into physical form. Physicality and the rules, norms, and laws of physicality are not based on divine unity, but on compartmentalized concepts of being. So, Love too becomes a physical concept when it's based upon a physical need. Now, a person may say that "a hug or smile is more than a physical act," and they're right! But then they're saying that "the physical act is creating something else!" Well, what is that something else? And they would say, "a feeling." That feeling is what many of us mistake as Love.

With a rather harsh comparison, heroin is injected or snorted, and in turn gives a person a feeling. Then that too must be Love, right? And then a person would clearly and concisely reject the idea, and claim, "no, this is a drug." Can you see the similarities? Our societal concept of Love is no different than a drug, because it's based upon fulfilling a need or desire for something we're missing. It's similar to when we express our understanding of chemistry between two people. You know the phrase, "they have great chemistry," or, "they didn't have chemistry." This interpretation shows that interaction between people is causing a chemical response. So when we have someone in our lives who we have "good chemistry" with, we find that they're actually sending emotional signals, which are activating enjoyable chemical releases within our bodies. Of course, these are the responses we're usually unable to activate on our own. This is just another example as to why our desire for fulfillment is based upon need-dependency of others.

Why are we not looking to solve this inward dilemma on our own? Why must there always be an outward source that fills us, even when it's clearly temporary? And the answer is simple… we're not educated on the matter! We're taught to be physically minded before we're taught to be emotional. Moreover, we're taught to be separate before we're taught that we're whole. When by the basic evidence of our biological development, we can clearly see that we're emotional beings before we're rational or physical. Furthermore, we're never separate! From this analysis, we come to the basic conclusion that we as a species are simply living backwards! In light of the divine experiment, and our original cosmic origin, we've gone in the opposite direction as was intended. Look at it this way… Because we're physically conscious beings, we should not be living to find the outward in life - we're already the furthest outward in the Universe - made as the measure of all things! How can we go further outward if we're already the furthest outward? Thus the furthest outward creation, should be searching far inward to know itself. And this is where we see ourselves living on just one side of a larger polarity, which must be experienced in its entirety to be understood. It's why the emotional realm seems so foreign to us, and why Love typically must be expressed in order to be felt - "we live with the feeling of being incomplete all of our lives!" Therefore, by our species choice to live solely for the outward, we only find half of the knowledge, truth, and Love that we desire. Yet when we go inward, far into our creation, we find a truth and a Love that's far more extraordinary, because of its wholeness or holiness.

The Divine Measure of the Anthropos: Unity Within

> "Here vigor failed the lofty fantasy:
>   But now was turning my desire and will,
>   Even as wheel that is equally moved,
>   By the Love which moves the Sun and other stars."
>   - "Dante" at the end of "The Paradisio"

Denying that every being exists in a polarity of male and female; electricity and magnetism; creativity and deconstruction; form and formlessness, and choosing only to exist on one end of the spectrum or the other, keeps us in a state of incompletion. By choosing not to look within ourselves and recognize our other half, we only find momentary experiences of completeness or Love in life. If we choose to remain this way, Love will always be associated with a fleeting feeling or a momentary expression for life's great beauty. True Love by this measure, is the Love that's experienced when the masculine and feminine aspects, within an individual, interlock and become one again. Through this junction, the expression of our inward sexual entanglement exists at the level of divine inspiration. Inspire of course referring to the nature of "in - spiraling," which realistically means energetic entanglement or a cosmic dance. It's a Birkeland current - a stream of energies entangled and spiraling between two connected points. So when Lovers dance together, their movements are not separate, but one. As a result of this, any and all relationships we have therein, while in this cosmic dance, are an expression of completeness from one point to the other - birth to death. Love always begins within, in this way, and spills out to embrace the whole of the world. So to experience holistic Love within oneself, unity must be expressed between the male and female; electric and magnetic principles, existing inside the individual.

By seeking to go inward to find one's own nature, we eventually find the aspects of ourselves that make us whole or holy. And we'll never have to go on searching for Love again, because that cosmic dance of completeness generates more than we can fathom.

Through this understanding, when we do see ourselves for who we are, we'll realize that we are and always have been the Love that makes the world go round. Human Beings are the product of a divine relationship, which is the representation of the oscillatory balance that resonates every star in the sky, and furthermore every galaxy in the Universe!

To bring this piece to a conclusion, we must understand that our individual being is seeking to find union between the two polar aspects of itself - the same as the greater cosmos. The view of our higher force and our lower force derives from examination of the holy trinity, which is merely an expression of the Universe - it's a hierarchy of being. The top point of the pyramid is the One Image or Great Mind of the Cosmos. The

Second point is the dual aspect or polar aspects of the One Image that can be described as male and female, positive and negative, or electricity and magnetism. These two forces of the One Image, when in union or cosmic dance, express themselves together to form matter, which gives us the third point... you! In fact, all matter that exists in the universe is the product of union between the two polar aspects of the One substance or All.

It can also be expressed as a diamond. At the top of the diamond we have the One Image. The two middle points are the polar opposites of the One Image, which are male and female or electricity and magnetism. The bottom point again, is you - the creation of the anthropos or divine matter brought forth by the Love of Creation.

So we must understand this Universe as something more than just a mystery that remains unresolved or endless, but rather from the nature of a divine marriage.

The two forces that merge together to create life, which are male and female - are the electric and magnetic principles of the Universe. When in union or yoga, they're co-creating, or taking part in the cosmic dance of time and space. The product of co-creation is how the One Image seeks to know itself. Therefore each and every one of us are not only the product of this co-creation, but are this co-creation in union with itself now. We're the electric and magnetic principles; male and female joined together - dancing the cosmic dance in the fullest, most beautiful expression of divine unification. Consciousness is therefore a careen of electromagnetism, continuing to find itself in each moment and through each movement. And so as a deduction, or further analysis, we must understand ourselves as participants in this cosmic boogie. We are ALL the One Image seeking to know itself through ecstatic play!

In order to find the One Image, we must first understand the two polar aspects of the cosmos joined together within us. That's moving your way up the ladder of knowing and reaching back into oneself(One Self).

When we look inward and witness the fusion of these two energies - delighting in ecstasy, we have to realize - on the opposite side of the observation deck exists the One source we're all seeking. And further, It's a mirror that reflects back at us. Inward is the way, because the divine intercourse or cosmic dance set forth by the One Image, who seeks to know and find itself now, exists as a complete energy within all of us. And because we're in polarity with the One Image, we have to remember that when we scale back and view this polarity - it's an expression of just One thing! It's all One! That's the hope and wish for this divine experiment; this anthropos, which we've loved and cherished through the wisdom of our Magna Mater; our Great Mother - the torrent of cosmic energy we call Gaia, Sophia, or the Earth... In time, humanity will fulfill the great wish and vision which has heralded the species as "the measure of all things." When that moment arrives, we'll laugh and dance our way into eternity, enveloped by the great Love of the cosmos.

Until then, we'll continue to wonder about the ifs, whens, whys and hows beyond our reason - delving further into the matter. After all, what would we do if we no longer had reason to wonder?

I'd imagine we'd simply wake up to another game - perpetuating Creation's need for further identification. So continue measuring, my friends. The spirit of all that is, and all that's in between seems to be right alongside us - doing the same! And because you seek to "know thyself," the bar continues to be set higher - rippling outward into the distant Universe.

It's your ripples which inspire the cosmos! So go big! Go far! Resonate that source within your heart. Illuminate the illusionary perception of darkness with your great luminosity. Show the world who you are, and fear no judgment! The judging world just needs time, as we all do - all things eventually come into their own.

Well... it's time to go now. I'll see you on down the stream, or, perhaps... further along to wherever it brings us. It's been a pleasure sharing the dream with you, my Love. The world continues to shine because you exist! So keep shining, keep breathing... And furthermore, know that not just the anthropos, but the entire Universe is continuing to fulfill its measure... through you - the Universal Being!

Conclusion

To answer the question that began our journey - "What kind of world do you wish to live in?" Which is the same as asking - "Who am I?" One would have to counter with another question... "If we're already the becoming and unbecoming nature of the cosmos - existing in ecstatic union with ourselves, why would we want to exist as anything else?"

"And the Wave continues inward and onward - oscillating the great poetry of the cosmos; a current of endless wisdom which perpetuates the greatest of truths... There's no end but a new beginning, which is always happening now - one moment of eternal expression, existing through you!"

www.ingramcontent.com/pod-product-compliance
Lightning Source LLC
Chambersburg PA
CBHW021545200526
45163CB00015B/1723